Quantum Computing Strategy

Quantum computing is not merely an incremental advancement in computing technology; it represents a fundamentally new paradigm, distinct from classical computing. Rooted in quantum mechanics, it introduces an entirely novel information theory. As a result, translating existing models, solution designs, and approaches to quantum computing is a complex, non-trivial task. This comprehensive book demystifies quantum concepts through accessible explanations, practical case studies, and real-world examples from industries such as aerospace, agriculture, automotive, chemicals, energy, finance, government, healthcare, manufacturing, supply chain, and telecommunications.

Blending a business perspective with a scientific rigor, this book is divided into two parts. The first part covers foundational technical concepts, including quantum mechanics principles that enable quantum technologies, key quantum algorithms, mathematical frameworks, quantum computing technologies, post-quantum cryptography, the types of problems quantum computers solve, and the technology's outlook. The second part focuses on practical applicability, presenting industry use cases, guidance on approaching quantum computing problems, mapping use cases to quantum computing, responsible quantum computing practices, and a roadmap for businesses preparing for quantum adoption. This structured approach equips readers with the knowledge and tools to effectively integrate quantum computing into their strategic planning.

Quantum Computing Strategy: Foundations and Applicability serves as an essential reference for technology enthusiasts, business leaders, policymakers, and educators seeking to understand the benefits quantum computing offers enterprises. Designed as a self-contained learning resource, it empowers readers to navigate the emerging quantum landscape confidently.

Elena Yndurain is a high-tech executive and professor with expertise in operationalizing innovation. She holds a Ph.D. and M.Sc. in Telematics Engineering focused on Artificial Intelligence, as well as an Executive MBA and a B.Sc. in Computer Science and Mathematics. With extensive international experience in consulting, technology, and banking, she has successfully launched numerous technology products. She was a pioneer in teaching quantum computing to business executives. Most recently, she served as the European CEO of a quantum computing software startup and is an active member of international advisory boards focused on technology governance.

Quantum Computing Strategy
Foundations and Applicability

Elena Yndurain

CRC Press
Taylor & Francis Group
Boca Raton London New York

CRC Press is an imprint of the
Taylor & Francis Group, an **informa** business
A CHAPMAN & HALL BOOK

Illustrations: Jorge Zeballos.

First edition published 2025
by CRC Press
2385 Executive Center Drive, Suite 320, Boca Raton, FL 33431

and by CRC Press
4 Park Square, Milton Park, Abingdon, Oxon, OX14 4RN

CRC Press is an imprint of Taylor & Francis Group, LLC

ISBN: 9781032299143 (hbk)
ISBN: 9781032275826 (pbk)
ISBN: 9781003302674 (ebk)

DOI: 10.1201/9781003302674

Typeset in Minion
by codeMantra

To my father—you would have enjoyed this book

Contents

Preface

THIS BOOK, QUANTUM COMPUTING *Strategy: Foundations and Applicability*, provides an overview of quantum technologies, focused on quantum computing, including specific use case applicability examples and concept definitions with the help of some simplification visualizations to facilitate the understanding from a high-level perspective.

This book marks not only the culmination of years of study and exploration but also a deeply personal journey into the world of quantum computing. This journey, enriched by a set of important experiences, has shaped not only my understanding of quantum computing but also my perspective on its strategic implications for business applicability.

MY JOURNEY INTO QUANTUM COMPUTING

My first encounter with quantum computing was in October 2000, when my father invited me to a talk by Spanish physicist Ignacio Cirac, who had defined the first trapped-ion quantum computer. In his talk, "Is it possible to build a quantum computer?", he explained the concept of quantum information, what a quantum computer is, and how we can build it. I was intrigued by this technology, which was so different from what had come before. In this conference, I was surprised to run into Jaime Botín, founder of Bankinter, a major Spanish commercial bank; he most likely had an intuition about how important quantum computing would be for the financial sector.

It was not until May 2016 that quantum computing came back into my life. At that time, IBM opened its quantum computer to the world via the cloud, providing access to early quantum computing prototypes to foster the creation of use cases and the development of future applications, as publicly stated by the company at that time.

I saw real quantum computers in a visit to IBM's quantum computer at their Thomas J. Watson Research Center in Yorktown Heights, NY, and to the Joint Quantum Institute (JQI) at the University of Maryland. I had the chance to get a first-hand overview of the computer in a lab tour by the researchers who were building the computer at both institutions – Jay Gambetta, Jerry Chow, and Mark Ritter from IBM as well as Prof. Luis Orozco and Prof. Mohammed Hafezi from JQI. At the time, none of these researchers were famous yet in the industry, and their language was candid and physicist-like, quite different from what there is today, which is an example of how quantum computing has gone to the forefront of business.

After these visits, one thing was clear to me: this technology was very different from other technologies. Quantum technologies are heavily based on quantum physics, and I decided to learn it so I could understand it better.

In 2017, I contacted Miguel Ángel Martín-Delgado, a full professor of theoretical physics at Complutense University of Madrid (UCM) and an expert in quantum information and error correction codes. He told me he believed that the next decade would be known as "the golden quantum twenties" and that expectations would be set for the so-called "second quantum revolution". In a series of meetings, he very generously taught me the foundations of quantum information and ignited my interest in self-study.

That same year, I added quantum computing to my own teaching repertoire. I created and taught a business overview course at IE Business School, targeted at master students from different programs. This was a first-of-its-kind initiative. It was one of the first courses where quantum computing was distilled into a format accessible to business students, focusing on its strategic implications rather than just its technical aspects. This course aimed to demystify quantum computing for future business leaders, enabling them to grasp how this technology could drive innovation, create new markets, and influence competitive industry dynamics.

In 2018, I got fully immersed in quantum computing by joining the early IBM Quantum team to work on business applicability and problem categorization. This work led me to become a quantum computing expert for startups, industry corporations, and academia, helping them become quantum ready at different levels.

A couple of years later, I created a more technical course for the Universidad Carlos III de Madrid (UC3M) Telecommunications Engineering and Cybersecurity program. The course, while more technical, was also designed with a focus on practical applications. This was a pioneering approach in that it provided engineering students with not just the theoretical foundations of quantum computing but also an understanding of how these principles could be applied to solve complex problems in their fields, such as secure telecommunications and data protection.

Until today, I continue to evangelize quantum computing through different universities, market research institutions, and corporations such as Stevens Institute of Technology, The University of Strasbourg, Arcano Economic Research, UC3M, Indra Sistemas, PG&E Corporation, and IE University, just to name some.

In my journey as both a learner and educator in the field of quantum computing, I have encountered a wide array of questions from students and executives alike. These recurring inquiries highlighted a common challenge: the need for a more accessible explanation of quantum computing concepts that would resonate with those outside the field of physics. Recognizing this, I embarked on a mission to demystify the complexities of quantum computing for a non-physicist audience.

The task at hand was not trivial. Quantum computing, with its roots deeply embedded in the principles of quantum mechanics, is inherently abstract and counterintuitive. The fundamental challenge lay in translating these intricate theories into concepts that are intuitive and relatable. To address this, I started by identifying the core elements of quantum computing that most captivated the curiosity of newcomers, regardless of their background or level of expertise. These ranged from the basic principles, like superposition and entanglement, to more advanced topics such as quantum algorithms and their potential applications.

One of the most common requests from learners, both in academic settings and the industry, was for a simplified visualization of quantum mechanics' foundational concepts. They sought to grasp not only how these principles differentiate quantum computing from classical computing but also the unique advantages and potential that quantum computing holds. To meet this need, I employed a variety of educational tools and analogies that resonate with everyday experiences, making the abstract concepts more tangible. For instance, I likened the principle of superposition to the multiple potential states of a spinning coin before it lands, offering a glimpse into the quantum world's probabilistic nature.

Moreover, there was a clear demand for real-world examples that illustrate the practical applications of quantum computing. Learners wanted to see beyond the theoretical framework to understand how this groundbreaking technology could address complex challenges across various industries. In response, I curated a collection of case studies and potential use cases, from drug discovery and material science to optimization problems and cryptography. These examples not only showcased the broad applicability of quantum computing but also highlighted its potential to revolutionize industries by solving problems that are currently intractable for classical computers.

STRUCTURE OF THIS BOOK

This book is split into two sections. The first section lays out the key concepts of quantum computing to ensure we understand this fascinating yet counterintuitive technology. It defines what makes quantum computing so special to set the seed for the *foundation* for the rest of this book. It covers the definition of algorithms explaining their key advantages and the underlying concepts in a high-level business-oriented way so that anyone can understand them. The second section covers the different quantum computing solutions providing *applicability* examples for different industries. It explains the impact of this technology from a business angle and how to get quantum ready from a corporation point of view.

HOW TO USE THIS BOOK

This book is designed to be entirely self-contained, meaning that it does not rely on external resources, additional texts, or prior extensive knowledge from the reader to provide a comprehensive understanding of quantum computing.

This approach ensures that readers can access and understand the full scope of quantum computing and its applications independently. The text includes a variety of real-world examples that demonstrate how quantum computing is applied in different industries and scientific research. These examples help to bridge the gap between abstract theoretical concepts and practical applications, making the material more accessible and engaging.

Additionally, this book incorporates numerous visualizations and illustrative diagrams. These visual aids are specifically designed to help clarify complex concepts and provide intuitive insights into the mechanics of quantum systems and algorithms. By using these visual tools, this book aids readers in grasping sophisticated ideas more effectively, enhancing their overall learning experience.

Overall, this book is meticulously organized to guide readers through the special world of quantum computing, from the basic principles to the more advanced topics, without the need for additional support. It is designed to bridge the gap between abstract quantum theory and practical computational strategies. I hope that this work will help students and professionals who are interested in new technologies in their learning process.

Acknowledgments

T HERE ARE MANY PEOPLE who have helped me with this book, directly or indirectly, through specific discussions about this book's topics, discussions of quantum computing in general, or simply with moral encouragement in difficult times.

Many generous people helped me peer review the content of my manuscript, brainstorming on industry analysis, use case challenges, quantum computing impact, or underlying quantum information concepts. In particular, the following people took time from their busy schedules to discuss or review some of the topics I was analyzing in the book on industry business challenges or quantum technology concepts.

They are listed here alphabetically by first name: Alejandro Peña Galiano, aerospace engineer; Alistair Nolan, OECD; Andrea Rodríguez-Blanco, post-doc at the University of California Berkeley; Antonio J. Montero Sánchez, telecommunications expert at MASMOVIL Group (now MASORANGE); Bob Sutor, ex-IBM research executive; Cristina Sánchez Muñoz, expert in automotive industry, manufacturing, and digitalization; Daniel Egger, quantum computing expert; David Muñoz, quantum chemistry expert; Emma Wang, expert in energy, OECD; Escolástico Sánchez, head of Quantum Computing, BBVA; Ewan Munro, expert in quantum information, COO Entropica Labs; Francisco Martín-Fernández, STSM at IBM Quantum; Hector Guerrero, expert in government and defense; Javier Echenique, expert in supply chain, CEO at IDLogistics Italy; José Ramón Tora, expert in financial services; Kai Watanabe, expert in quantum computing technology and business strategy; Kesha Sorathia, computational chemistry expert; Kumar Bhaskaran, IBM researcher; Laura Barea, expert in pharma at AbbVie; Lauri Toikka, quantum information researcher; Marwa Farag, quantum computing and quantum chemistry expert; Michael D. Cohen, technology strategy expert; Michele Grossi, coordinator quantum algorithm at CERN,

Miguel Ángel Martín-Delgado, quantum information expert, Universidad Complutense de Madrid; Ricardo Enriquez, Quantum Technologies for industry expert; Ricardo Olalla, expert in agriculture and digitalization, BOSCH; Rima Kasia Oueid, expert in government and quantum commercialization; Roman Orús Ikerbasque Quantum Information Professor and Multiverse Computing co-founder, Sharana Kariappa, optics engineer at Nu Quantum; Tom Lubinski, quantum programming software architect at Quantum Circuits Inc; and Veronica Fernandez-Mármol, quantum communications expert and CSIC tenured scientist.

I would like to give Tom Lubinski special thanks for spending additional time brainstorming and reviewing some key concepts of this book.

I also want to mention the assistance from IE Business School students who researched some of the information that I needed as a source of analysis: Rafael Cifuentes, Ashley Duffins, Ishita Kishore, Ramez Kabbani, Alex VanOeveren, and Marnix Van der Stighelen, with special mention to fellow student Abdullah Ruwaili from McKinsey & Company, who collaborated on the use case business impact analysis.

Apart from the people who have directly contributed to this book, there have been others who have indirectly done so, from whom I learned about different quantum technology topics in the different industry events that I have attended or watched online, such as the IEEE Quantum Week led by Hausi Muller, the Quantum Economic Development Consortium (QED-C) led by Celia Merzbacher, the Q2B conference led by Matt Johnson, and the Commercialising Quantum Global conference led by the Economist. In some of those events, apart from participating, I was also able to discuss quantum technologies with other experts. I had the opportunity to hear about the state-of-the-art of quantum technology from quantum computing companies, investors, industrial corporations, academics, researchers, and government representatives.

Of course, I want to extend my thanks to my publisher, Randi Slack, for her patience and support while I prepared the book's manuscript and to the Taylor & Francis Group for its interest in helping advance Quantum Computing adoption making it accessible to all audiences.

Finally, I again want to thank Prof. Miguel Ángel Martín-Delgado for suggesting and encouraging me to write this book to help link research to business.

Glossary

A

Aerodynamics: A branch of fluid dynamics that studies how gases interact with moving bodies to understand how objects move through the air.

Advanced Encryption Standard: A symmetric encryption algorithm widely used across the globe.

Air Cargo Load: Weighted, correctly loaded, and properly secured cargo in aircraft to prevent movement in flight.

Air Traffic: The aircraft, or the number of aircraft, flying in an area or along a route

Annealer: A special type of quantum computer designed to solve optimization problems.

Autonomous Vehicles: Self-driving cars capable of sensing their environment using sensors and moving safely with little or no human input.

B

Bell State: A specific type of quantum state that involves two qubits in a maximally entangled state, named after the physicist John Stewart Bell.

Binding Affinity: Strength with which a drug binds to its target involved in a disease process in the body, usually a protein such as an enzyme or receptor.

Bloch Sphere: A geometrical representation of the pure state space of a two-level quantum mechanical system (qubit), named after the physicist Felix Bloch.

Business Advantage: Anything that gives a company an edge over its competitors, in terms of cost advantage, attracts more customers, or grows its market share.

C

Carbon Capture Utilization: Techniques to use a harmful waste product (CO_2, which contributes to climate change) in industrial processes by capturing carbon dioxide emissions from generation sources.

Central Processing Unit: It is the most important processor in a computer, it has electronic circuitry that executes instructions of a computer program.

Catalyst: A substance that increases the rate of a chemical reaction without itself undergoing any permanent chemical change.

Coherence: Ability of a quantum state to maintain its entanglement and superposition in the face of interactions and the effects of thermalization.

Cryptographic Agility: Ability of security hardware to update an algorithm without the need to rewrite applications or deploy new hardware systems.

CRYSTALS-Kyber: A lattice-based cryptographic algorithm.

D

Decoherence: The process by which quantum properties decay over time.

Derivatives: Financial contracts whose value is dependent on an underlying asset, group of assets, or benchmark.

Drug Discovery: The process of identifying a new medication by finding compounds that can effectively treat or prevent diseases.

Dynamic Routing: A networking technique used to find the best path for data to travel across a network that automatically adjusts the paths.

E

Electric Battery: A source of power that converts chemical energy into electrical energy through an electrochemical reaction and is rechargeable.

Elliptic Curve Cryptography: Form of public-key cryptography based on the algebraic structure of elliptic curves over finite fields.

European Telecommunications Standards Institute (ETSI): Independent, not-for-profit, standardization organization in the field of information and communications.

F

Fault-Tolerant Quantum Computer (FTQC): An error-corrected quantum computing system designed to perform accurate computations.

Flight Climb: The operation of increasing the altitude of an aircraft.

Fraud: Deceptiveness or misrepresentations to gain a financial or personal advantage, or to cause harm to others.

G

Genomic Analysis: Study of DNA sequencing, structural variation, gene expression, or regulatory and functional element annotation at a genomic scale.

Gross domestic product (GDP): A monetary measure of the market value of all the final goods and services produced and rendered in a time period by a country.

H

Hamiltonian: A fundamental concept of quantum mechanics that represents the total energy of a system kinetic and potential.

High-Performance Computing (HPC): Use of supercomputers and computer clusters to solve advanced computation problems.

Hilbert Space: A mathematical framework for representing and manipulating quantum states, named after the mathematician David Hilbert.

I

Information and Communication Technology (ICT): Infrastructure and components that enable computing to manage information and facilitate communication.

Irregular Operations (IROPs): Disruptions causing customers not to be able to use the flight(s) ticketed.

International Organization for Standardization (ISO): Independent, non-governmental, international standard development organization composed of representatives from the national standards organizations of member countries, HQ in Europe.

K

Key Encapsulation Mechanism: A cryptographic protocol used in securing communication between two parties by encapsulating a secret key.

KET: A column vector in the space of quantum states to describe the state of a quantum system.

L ˙

Light weighting Design: Constructing cars and trucks with lighter designs to enhance fuel efficiency and improve handling.

Linear Algebra: A branch of mathematics about linear equations fundamental to quantum computing to describe qubit states and perform quantum operations.

Logical Qubit: A combination of qubits that work together to perform computations with a lower error rate.

M

Machine Learning: A branch of artificial intelligence that focuses on using data and algorithms that are able to learn and adapt without following explicit instructions.

Material Compounds: Compounds made from two or more constituent materials with different physical or chemical properties.

Mean time between failure (MTBF): Measure of the reliability of a system or component used in the context of industrial or electronic system maintainability.

Molecular Energy States: All the energy forms that molecules can possess due to their atomic arrangements and movements, which exist at a specific energy level.

Multiple Input Multiple Output: A wireless communication technology that uses multiple antennas to increase channel capacity.

N

National Institute of Standards and Technology (NIST): An agency of the US Department of Commerce with the mission to promote national innovation and industrial competitiveness.

National Security Agency/Central Security Service (NSA/CSS): Leads the U.S. Government in cryptology encompassing both signals intelligence and cybersecurity.

Noisy Intermediate-Scale Quantum (NISQ): A class of quantum computing devices that are currently in development and used with significant errors in them.

P

Post-quantum Cryptography: Cryptographic algorithms that are believed to be secure against an attack by a quantum computer, i.e., quantum-resistant cryptography.

Q

Quantum Advantage: The goal of quantum computers to outperform classical computers.

Quantum Algorithm: A step-by-step procedure, using the principles of quantum mechanics, to solve a problem or perform a task.

Quantum Amplitude Estimation: An algorithm to determine the probability of measuring a specific outcome in a stochastic system.

Quantum Annealing: A quantum algorithm for solving optimization problems by finding low-energy states of a system.

Quantum Circuits: The basic structure for processing quantum information, consisting of qubits and quantum gates.

Quantum Communications: A technology using devices that transmit un-hackable information.

Quantum Computers: Devices that use quantum mechanics for computation.

Quantum Cryptography: An encryption method that uses quantum mechanical properties to secure data.

Quantum Entanglement: A phenomenon where quantum particles become interconnected and the state of one instantly influences the state of another.

Quantum Error Correction: A set of techniques to protect quantum information from errors due to decoherence and other quantum noise.

Quantum Fourier Transform: A linear transformation used in many quantum algorithms to map quantum states and amplitudes.

Quantum Gates: Physical operations performed on qubits to change their states, analogous to logic gates in computer science.

Quantum Inspired: Use of computational methods inspired by the principles of quantum mechanics.

Quantum Interference: A quantum mechanics phenomenon where the probability amplitudes of quantum states overlap and combine.

Quantum Key Distribution: A secure communication method that implements a cryptographic protocol through quantum mechanics to distribute and share keys.

Quantum Mechanics: A physical theory that describes the physical properties of nature at the scale of atoms and subatomic particles.

Quantum Noise: A measured vibration that can lead to inaccuracy.

Quantum Parallelism: The possibility of performing a large number of calculations simultaneously by applying quantum properties.

Quantum Processing Unit: The core component of a quantum computer, and it is the chip where the computation takes place.

Quantum Random Number Generators: Hardware components used for generating secure and unpredictable codes used for cryptography.

Quantum Secure Cryptography: Cryptographic techniques designed to be secure against quantum computing attacks.

Quantum Sensors: Devices that detect physical properties with high sensitivity.

Quantum Superposition: A principle where a quantum system can be in multiple states at the same time.

Quantum Support Vector Machine: Quantum version of machine learning algorithms that maps classical data into a high-dimensional quantum feature space.

Quantum Teleportation: A process of transferring quantum information from one location to another, using entanglement.

Qubit: The basic unit of quantum information, analogous to a bit in classical computing.

R

Reservoir: An oil and gas reservoir is a rock formation in which oil and natural gas have accumulated and are trapped in connected pore spaces of rock.

Rivest-Shamir-Adleman: Public-key cryptosystem used for secure data transmission.

S

Schrödinger Equation: Fundamental equation in quantum mechanics that describes how the quantum state of a physical system changes over time measuring energy levels, named after the physicist Erwin Schrödinger.

Speedup: Process that reduces the number of steps to solve a problem either significantly (exponential) or slightly (polynomial).

Statistical Process Control: The use of statistical techniques to control a process or production method to help improve quality in performance.

Sustainable Development Goals: The 17 world goals part of the 2030 Agenda adopted by all United Nations members in 2015 that highlight the connections between the environmental, social, and economic aspects of sustainable development.

Supply Chain: The network of individuals, organizations, resources, activities, and technology involved in the creation and sale of a product.

T

Technology Adoption Cycle: A sociological model that describes the acceptance of a new product based on the demographic and psychological characteristics.

Thermodynamics: The relationship between heat, work, temperature, and energy.

Threshold Theorem: A framework for quantifying the fault tolerance level to effectively perform quantum computations.

Technology Readiness Level: A scale that assesses the maturity of technologies, ranging from initial concepts to fully operational systems.

Tensor Product: A mathematical operation combining two or more vectors or matrices, used in quantum circuits to combine qubits with quantum operators.

Traveling Salesman Problem: A classic problem in a combinatorial optimization problem that seeks to find the shortest path to visit different cities.

Tunneling: A quantum mechanical phenomenon where a particle passes through a potential barrier despite having less energy than the height of the barrier.

Turbines: Machines for producing continuous power in which a wheel or rotor, fitted with vanes, is made to revolve by a fast-moving flow of different fluids.

U

Unit Commitment Problem: A problem that aims to minimize the total cost of electric power generation in a specific period, balancing and scheduling the generating units.

Unitary Evolution: Refers to the evolution of a time-dependent quantum system that preserves the norm of the state vector, ensuring probabilities sum to one.

Uncertainty Principle: Certain pairs of physical properties, like position and momentum, cannot both be precisely measured at the same time.

V

Vehicle Routing: A problem that seeks to find optimal routes for multiple vehicles visiting a set of locations.

Variational Quantum Algorithms: Hybrid algorithms that use both classical and quantum computing resources to find approximate solutions to problems.

Acronyms

AA	amplitude amplification
AES	advanced encryption standard
AI	artificial intelligence
ATM	automated teller machine
BCG	Boston Consulting Group
CAGR	compound annual growth rate
CCUS	carbon capture, utilization, and storage
CEO	chief executive officer
CISA	Cybersecurity and Infrastructure Security Agency
CPU	central processing unit
CT	computed tomography
ECC	Elliptic Curve Cryptography
ETSI	European Telecommunications Standards Institute
FAA	Federal Aviation Administration
FTQC	fault-tolerant quantum computer
GDP	gross domestic product
GPS	global positioning system
HHL	Harrow–Hassidim–Lloyd
ICT	information and communications technology
IEEE	Institute of Electrical and Electronics Engineers
IMF	International Monetary Fund
IP	internet protocol; intellectual property
IP-VPN	internet protocol-virtual private network
IROPs	irregular operations
ISO	International Organization for Standardization
IT	information technology
ITU	International Telecommunications Union
JQI	Joint Quantum Institute
KEM	key encapsulation mechanism

KYC	know your customer
MIMO	multiple-input, multiple-output
mK	millikelvin
MPLS	multiprotocol label switching
MRI	magnetic resonance imaging
MTBF	mean time between failure
NSA	National Security Agency
NASA	National Aeronautics and Space Administration
NGB	National Guard Bureau
NISQ	noisy intermediate-scale quantum
NIST US	Department of Commerce's National Institute of Standards and Technology
OECD	Organization for Economic Co-operation and Development
OEM	original equipment manufacturer
OPEX	operational expenditures
PET	positron emission tomography
PPPs	plant protection pesticides
PQC	post-quantum cryptography
QAE	quantum amplitude estimation
QAOA	quantum approximate optimization algorithm
QDE	quantum differential equation
QEC	quantum error correction
QED-C	quantum economic development consortium
QFT	quantum Fourier transform
QML	quantum machine learning
QNN	quantum neural network
QPE	quantum phase estimation
QPU	quantum processing unit
QUBO	quadratic unconstrained binary optimization
QRNGs	quantum random number generators
QSVM	quantum support vector machine
PII	personally identifiable information
R&D	research and development
RAN	radio access network
RSA	Rivest-Shamir-Adleman
SDG	sustainable development goals
SDK	software development kit
SD-WAN	software-defined wide area network
SPC	statistical process control

STSM	senior technical staff member
TRL	technology readiness level
TSP	traveling salesman problem
UC3M	Universidad Carlos III de Madrid
UCP	unit commitment problem
VQA	variational quantum algorithms
VQE	variational quantum eigensolver
VR	vehicle routing
UN	United Nations
WEF	World Economic Forum

Introduction

W ELCOME TO *QUANTUM COMPUTING Strategy: Foundations and Applicability*, a journey into one of the most enigmatic and transformative technologies of our age. In these pages, we will explore the world of quantum computing, a field that challenges our traditional notions of computation and information processing. This book doesn't require advanced mathematical or physics knowledge. It is written for a broad audience, including technology enthusiasts, professionals exploring quantum computing applications in their fields, and policymakers seeking to understand the strategic importance of this emerging technology.

This book covers all necessary technical foundations for understanding what quantum computing is; contextualizes it in the larger quantum technology scheme; explains its key underlying algorithms; defines the problem areas that will benefit most from this technology; and analyzes how to choose an appropriate problem to be solved with this technology, how to approach its development, how to link problem areas with industrial use cases, how to work in a hybrid quantum–classical environment, and how to integrate quantum computing with existing technologies.

The target audience for this book is curious professionals who evaluate technologies and are considering quantum computing, listed below alphabetically:

- Anyone interested in gaining an intuition on how quantum computing works, its use case potential, and an overview of the whole ecosystem.

DOI: 10.1201/9781003302674-1

- Board members who want to understand tech implications and opportunities.

- Business leads who need to understand the benefits of quantum technology.

- DevOps architects who define how to integrate technologies.

- Financial advisors making decisions on high-tech investments.

- Graduate students who want to learn quantum computing.

- Instructors who need to teach this topic to non-STEM students.

- Journalists specialized in technology topics and the latest innovations.

- Product owners who need to determine the applicability of use cases.

- Regulators and policymakers who need to understand the power of quantum technology to assess how to regulate it.

- Software developers who need to obtain a technology and use case applicability overview.

Some of the concepts covered in this book are the peculiar properties of quantum bits (qubits), how they can store multiple states (superposition), how they interact through highly correlated behavior (entanglement), how the algorithms are different from existing ones, how they outperform existing computational systems, and how to categorize the type of problems that they solve including quantum security, the other side of the coin of the technology use, and finally a guide on technology adoption roadmaps.

Although this book defines quantum technologies in general, it is focused on quantum computing and on some of the applicability use cases that industrial corporations are testing. Corporations are thinking about problems that are difficult to solve, cannot be solved quickly, or cannot be properly approached, and this is what motivates them to test quantum computing.

Some examples of quantum computing applicability for business areas covered in this book include the following:

- Aerospace companies are testing air analysis when flying, monitoring operational uncertainties during flight, and material design.

- Agriculture farmers are testing how to improve the quantity and quality of crops and become more sustainable.

- Automotive companies are testing applications in battery design, routing, and vehicle production process.

- Chemical companies are testing catalytic reactions, electrochemical effects, and correlated systems in computational chemistry.

- Energy companies are testing energy resource optimization, reduction of carbon emissions, and management of power generation networks.

- Financial entities are testing applications for risk analysis in portfolio management, product pricing, and fraud detection.

- Government agencies are testing how to enhance country security, city management, and sustainability.

- Healthcare pharmaceutical companies are testing how to improve patient diagnosis and medical treatments.

- Manufacturing companies are testing how to optimize production lines and development of new materials.

- Supply chain and logistics companies are testing how to optimize demand and transportation management.

- Telecom companies are testing how to improve networks from planning to operations to handling multiple devices.

This book provides insights into how businesses, governments, and individuals can prepare for the upcoming quantum era, highlighting the challenges and opportunities that lie ahead. The purpose of this book is twofold.

First, it serves as a primer on the core principles that underpin quantum computing, covered in the first section. Second, this book explores the strategic implications of quantum computing on different industries, covered in the second section.

This book is intended for those who want to understand the technological benefits of quantum computing, the problems it solves, the areas it affects, and its impact on current operations. The goal of this book is to serve as a "handbook" and as a guide and reference. As such, it can be read sequentially, building knowledge from chapter to chapter, or as a reference guide, where the reader can consult a specific section to understand a problem described for a particular field or industry, the applicability of a use case, how a specific algorithm works, or a detail about the technology stack.

As you turn these pages, I invite you to join me on a journey of discovery, exploring the strange yet fascinating world of quantum computing. It is a journey that will not only expand your understanding but also challenge you to think about the future of technology and its impact on our world.

SECTION I

Foundations

Our world is quantum, as it is made of atoms; all the things that surround us are made of atoms, from cars to trees to our bodies. Therefore, it makes sense to apply the laws that govern quantum objects to help us understand the world. This is achieved through quantum mechanics, a physical theory that describes the fundamental properties of nature at the scale of atoms and subatomic particles. Its principles underpin many everyday phenomena that are vital to our lives.

Nobel laureate Richard Feynman explained in the 1980s that our world is a quantum world, and hence, we must be able to use quantum systems to help us better understand it. He proposed to create a computer that would take advantage of the effects of quantum mechanics with the goal of performing challenging tasks that no classical computer can do, in particular, simulating the behavior of complex quantum systems that are out of reach for classical machines. This computer would combine quantum physics, computer science, and information theory as envisioned by the pioneering theoretical physicist David Deutsch, by extending the Turing machine into the quantum world.

In the early 20th century, we saw the **first quantum revolution**, when the fundamental principles of quantum mechanics were discovered and developed, which changed our understanding of physics and the nature of reality, leading to numerous scientific and technological advancements.

DOI: 10.1201/9781003302674-2

This laid the groundwork for technologies such as transistors, semiconductors, lasers, nuclear energy, and atomic clocks used for the following:

- GPS in our phones works through transitions in quantum states.

- Medical magnetic resonance devices are powered by the properties of atomic nuclei.

- Supermarket barcode laser readers are enabled through photon and electron stimulation.

- Transistor operations rely on the behavior of electrons at a quantum level.

- Optical fiber communications operate by modulating light, altering its intensity or frequency.

Quantum mechanics was used to understand and describe quantum phenomena to actively harness, control, and apply these phenomena in technology.

In the late 20th century, there was a **second quantum revolution**, enabling us to apply the principles of quantum mechanics to manipulate individual atoms and electrons and to examine and shape our world through a quantum lens.

Once scientists knew the rules of quantum mechanics and built devices that followed those rules, they used them to develop new technologies. This engineering approach allowed them to manipulate quantum behaviors towards specific outcomes.

It started a transformative era, in which we have the technological prowess to manipulate, control, and exploit individual quantum systems for practical applications relying on direct control and use of quantum phenomena. It holds promise for revolutionizing fields ranging from computing and telecommunications to metrology and materials science.

Quantum states can be controlled with high precision, because with the second quantum revolution, atoms can now be cooled, stabilized, controlled, and manipulated.

The advancements in quantum technologies have broad implications across numerous fields, including computer science, materials science, biology, engineering, and many industries, which we will review in the applicability section of this book.

The second quantum revolution gave birth to quantum technology, which covers different technical areas that we describe in detail in the next page:

1. Sensors measure physical quantities with more precision.

2. Communications enable secure communication avoiding eavesdropping.

3. Computers perform certain calculations much more efficiently.

Before we define each of the concepts, it is important to highlight the field of Quantum Metrology, which focuses on achieving high-resolution and highly sensitive measurements of physical parameters through the application of quantum theory to describe physical systems. This area holds the potential to enhance measurement precision beyond what is achievable within a classical framework, serving as a foundational theoretical model for quantum sensing.

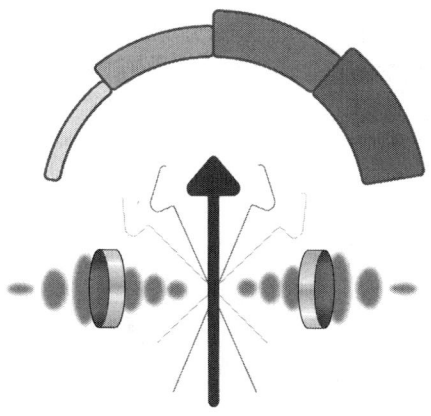

A quantum sensor is a device that works by detecting variations in microgravity using the principles of quantum physics by sensing changes in motion and in electric and magnetic fields. The analyzed data are collected at the atomic level. Such a sensor can detect a wide range of tiny signals from the world around us without the limiting environmental factors, such as vibration, which affect current gravity sensors.

Potential impact: Quantum sensors will be able to see the invisible world, which will help us to predict, for example, natural disasters or climate change. They measure different physical properties like temperature, magnetic field, and rotation, with extreme sensitivity.

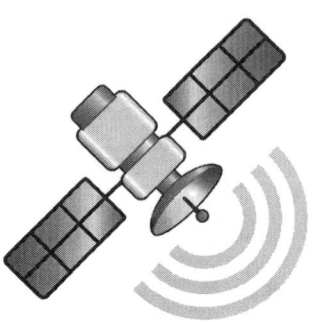

Quantum communication information is represented as a series of quantum states instead of bit strings. The channel maps the quantum state of the input to the quantum state of the output. It transforms alphabets into coded quantum states, and it finds the result by measuring quantum states and decoding them into the alphabet. It enables high capacity and speed as well as error correction.

Potential impact: Quantum communications allow unbreakable encryption and strengthen data cybersecurity. In this sense, they are designed to transfer quantum information between distant locations using cryptography.

Quantum computers represent information utilizing atomic phenomena at the microscopic scale and applying systems based on quantum mechanics; their hardware and software are based on quantum mechanics. They are universal processors that are expected to be able to solve problems that are unsolvable by supercomputers.

Potential impact: Quantum computers address complex global sustainability problems. They manipulate the energy states of atoms to process information and solve calculations more efficiently, through advanced computation, bridging the fields of physics and computer science.

CHAPTER 1

Key Concepts

T HE JOURNEY FROM QUANTUM mechanics to quantum computing spans over a century, starting in 1900 with Planck's proposal of quantized energy to explain black-body radiation. This pivotal development laid the foundation for quantum mechanics, a branch of physics that describes the behavior of energy and matter at the atomic level using probabilistic and non-intuitive aspects.

This field rapidly evolved with several significant contributions, such as Schrödinger's wave mechanics in 1926, enabling particle behavior prediction; Heisenberg's Uncertainty Principle in 1927, a fundamental measurement theory for quantum mechanics; and Dirac's comprehensive framework on quantum mechanics in 1930, introducing the bra-ket notation to describe quantum states.

The conceptual leap to quantum computing began in the 1980s. First with Paul Benioff's theoretical model, followed by Feynman's 1981 proposal of a quantum-based computer designed for simulating quantum systems. David Deutsch then formulated the concept of a universal quantum computer in 1985, introducing the ideas of quantum superposition and entanglement properties, which officially gave rise to the field of quantum computing.

Throughout the 1980s, foundational principles like quantum key distribution and quantum teleportation were defined by Charles Bennett and Gilles Brassard. These developments led to Peter Shor's factoring algorithm and Lov Grover's search algorithm to show quantum speedup

DOI: 10.1201/9781003302674-3

in the 1990s. Additionally, breakthroughs in communications were made with the quantum key distribution (QKD) protocols and quantum teleportation, forming the foundational principles for quantum information theory.

The 21st century marked the transition from theoretical to practical applications in quantum computing. The key milestones include the introduction of the D-Wave One in 2011, the first commercial quantum computer; IBM's launch of a cloud-based quantum computer in 2016, and China's deployment of its first quantum communication satellite using QKD in 2017; and Google's demonstration of "quantum supremacy" in 2019, when the company announced its Sycamore 53-qubit quantum computer completed a long task in only 200 seconds.

1.1 QUANTUM INFORMATION

Quantum information refers to the representation, manipulation, and communication of information based on the principles of quantum mechanics. This emerging field integrates science and technology, combining physics, mathematics, computer science, and engineering to understand certain fundamental laws of physics and enhance the acquisition, transmission, and processing of information. Quantum computers store and process encoded information in qubits. Quantum information enables the execution of tasks that would be either impossible or very difficult in a classical world, which we will review in this book.

This new discipline aims to understand how information is processed and transmitted using quantum mechanical principles. By integrating quantum mechanics with information and computation theory, it advances theories, algorithms, and technologies, enabling us to exceed the boundaries of traditional computing.

The shift toward quantum information began as scientists explored the peculiar behaviors of particles at the quantum level. Unlike classical information, which is encoded in binary digits (bits) that can be either 0 or 1, quantum information is encoded in quantum bits or qubits. A **qubit or quantum bit**, the equivalent of the classical bit, is fundamental to quantum computation and serves as the physical carrier of quantum information. Figure 1.1, Figure 1.2, and Figure 1.3 show the qubit concept.

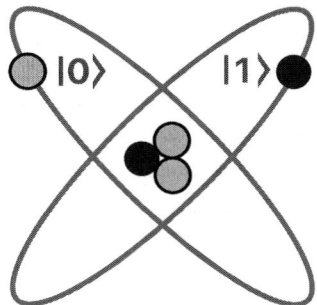

FIGURE 1.1 Atom's electronic levels.

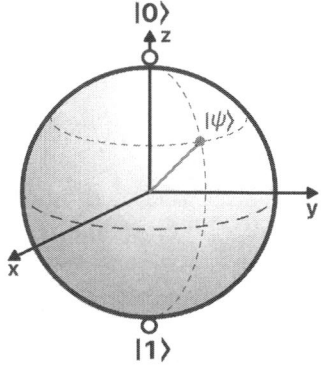

FIGURE 1.2 Bloch spin sphere.

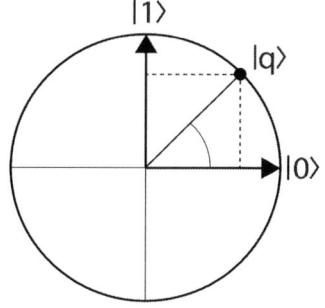

FIGURE 1.3 Hilbert space.

Qubits, just like bits, can have special states |0 or |1, which are mathematically denoted using Dirac brackets|, known as kets. This means that |0 +|1 is not like the standard 0+1. For example, a qubit is represented through a particle's spin state.

Qubits behave like atoms, performing calculations by manipulating how an electron spins on its axis; one direction represents the '0' state, and the opposite direction represents the '1' state. An electron can exist in two states, ground or excited.

Qubits are geometrically represented on a three-dimensional sphere, known as Bloch sphere, that helps visualize a qubit and its operation through a "state vector" representing different states of the qubit.

This vector is represented by a linear combination of states $|\psi = \alpha|0 + \beta|1$, where α and β are complex numbers known as probability amplitudes. It operates within Hilbert space, a mathematical representation used to perform operations describing physical systems through "wavefunctions" that represent the possible states of the qubit.

Qubits are time-dependent, two-energy level quantum mechanical systems whose evolution is governed by the Schrödinger wave equation.

It describes the transformation of the input state to an output state by how the state vector evolves over time. The qubit resides on the complex circle in the Hilbert space encompassing all possible orientations of its states. This qubit's position on this sphere is defined by two angles, θ and φ, used to measure its movement.

1.2 QUANTUM PROPERTIES

The processing power that enables quantum computers to surpass classical computers stems from the computer's core: the qubits, and their properties. Qubits provide **exponential calculating power** because they can exist in two states and can store up to 2^n values, where n is the number of qubits

Therefore, if we have two qubits, we can store four values; with three qubits, we can store eight values; with 20 qubits, we can reach 1 million values; and so on. This continues until massive amounts of data are stored.

In Table 1.1, we can see how it works.

TABLE 1.1	Illustration of Exponential Power of Qubits	
N	**Classical Bits ($2 \times N$)**	**Quantum Bits (2^N)**
2	$2 \times 2 = 4$	$2^2 = 4$
3	$2 \times 3 = 6$	$2^3 = 8$
4	$2 \times 4 = 8$	$2^4 = 16$
...		
100	$2 \times 100 = 200$	$2^{100} = 1,267,650,600,228,229,401,$ $496,703,205,376$

In addition, qubits can function as a single "block" when entangled. Consequently, any action performed on one qubit affects the entire entangled block, and reading information on one provides information on the rest. This means we can have a block of 2^n values at our fingertips to perform massive parallel calculations.

So, for an 8-qubit system, we can calculate 28 operations of 256 amplitudes simultaneously. The classical equivalent would need 256 processors (or cores), each updating one value in parallel. This phenomenon, known as quantum parallelism, provides exponential speedup, enabled by combining the superposition principle and constructive interference of their amplitudes to compute solutions.

It is worth noting that quantum parallelism is advantageous primarily for certain types of complex problems that involve high dimensional data, rapid growth, or extensive calculations. The processing power of this technology derives from a set of four key mechanical properties, namely, superposition, entanglement, interference, and teleportation, that collectively bring exponential power. We will explore each of them separately.

- **Superposition** refers to the property that a qubit can take several values: 1, 0, and all intermediate values.

- Figure 1.4 illustrates a conceptual visualization of this property by using the analogy of a coin.

FIGURE 1.4   Superposition concept visualization.

- The coin has two sides, heads and tails. However, when spinning, it simultaneously displays both head and tail values transitioning from a circle to a sphere. We do not know its value until we stop the spinning. It is only when we stop the spinning, and "observe" the coin, that we will see what side it has landed on. This is what Schrödinger's famous observation principle describes. It amounts to measuring the qubit to retrieve the information stored within it. Schrödinger provided an example in which he placed a cat in a sealed box with a mechanism that can kill the cat based on the decay of a radioactive atom. The cat is simultaneously alive and dead until the box is opened and observed.

- **Entanglement** refers to the fact that when particles interact, their states become correlated. Once one qubit is measured, the second one is instantly determined. Particles become interconnected in such a way that the state of one (no matter how far apart) instantaneously affects the state of another. It is enabled by quantum phenomena.

- Figure 1.5 provides a conceptual visualization of this property by using the analogy of two magnets.

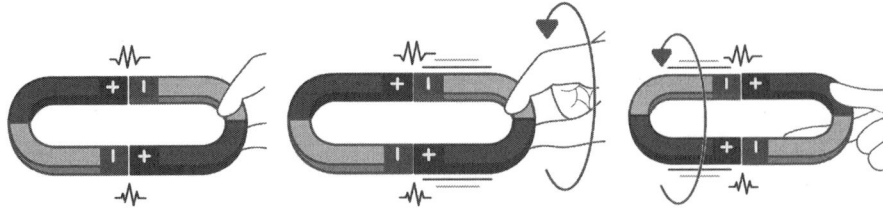

FIGURE 1.5   Entanglement concept visualization.

- Let us consider two magnets that are attracted to each other. When we rotate one of them, the other automatically switches. Therefore, the change in the behavior of one affects the other instantaneously.

- Entanglement is what Albert Einstein referred to as "spooky action at a distance". He used this term to express his skepticism and discomfort since it seemed to contradict the principles of locality and causality in classical physics. The term "spooky" came from how entangled particles seem to "communicate" over long distances without any known mechanism or signal passing between them, defying the intuitions about how the world operates.

- **Interference** occurs when there is more than one way to obtain a particular result, as particles can follow multiple paths simultaneously. This interference is enabled by the superposition quantum mechanical phenomena.

- Figure 1.6 provides a conceptual visualization using the analogy of how water waves expand.

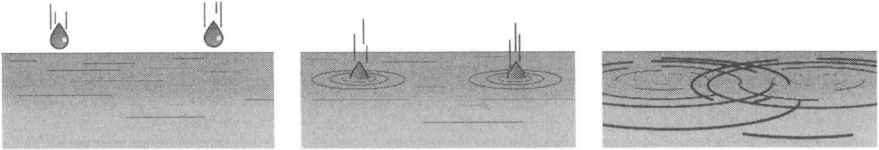

FIGURE 1.6   Interference concept visualization.

- Waves have a certain interference pattern when they interact; the ripples boost or cancel their peaks and troughs causing the wave to exist in two places at once. This behavior allows us to measure superposition states. This concept was demonstrated in the double-slit experiment. The rationale of quantum computing is to enhance constructive interference of states to amplify the probability of finding the desired output of a quantum computation.

- **Teleportation** is the process of transporting quantum information in a qubit from one location to another without physically moving the qubits-associated particles, through entangled states.

- Figure 1.7 shows a conceptual visualization by using the analogy of a fax machine.

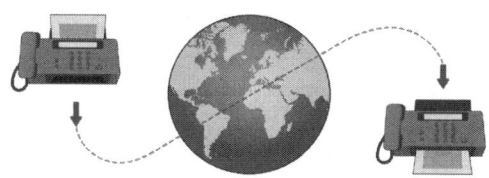

FIGURE 1.7   Teleportation concept visualization.

- A classical fax machine scans a document and transfers its content through a signal to another fax machine, which then prints out the document. This way the transfer is performed by copying the classical state, known to be a Bell state that represents a two-bit entanglement.

- Quantum teleportation is enabled by entangled states. It allows us to transfer the quantum state between two parties at the expense of destroying the information in the initial state because cloning is not possible, unlike in classical computing. We cannot create a duplicate of a quantum bit in an unknown state without perturbing the original. This is how teleportation is secure against cyberattacks, unlike classical faxing.

- It is important to reinforce the notion that quantum teleportation does not transmit the original particle itself, nor does it allow for faster-than-light communication, as it is sometimes misunderstood. The process essentially "teleports" the quantum state of a particle, not the particle itself. Additionally, it must be noted that for teleportation to occur, a classical channel is required to transmit the result of the qubits measurement from the sender to the receiver. The original state is destroyed in the sending location and recreated in the receiving location, adhering to the quantum no-cloning theorem, which states that it is impossible to create an identical copy of an arbitrary unknown quantum state.

1.3 MATHEMATICAL DEFINITIONS

Linear algebra, a branch of mathematics, is used for operations and linear transformations or maps. It is the fundamental language of quantum computing and quantum information, used to describe qubit states and perform quantum operations. It is the language of matrices and vectors. Some of its key concepts are defined below:

- **Complex numbers** (α or β) are elements of a number system that extends the real numbers and are used to represent and manipulate quantum states, $\alpha|0\rangle + \beta|1\rangle$

- **Matrices** are a formation or arrangement of complex numbers Vij, used to represent linear maps and allow explicit computations in linear algebra; they are used to represent quantum gates, which have a size of $m \times n$.

$$\begin{bmatrix} V11 & \cdots & V1n \\ \vdots & \ddots & \vdots \\ Vm1 & \cdots & Vmn \end{bmatrix}$$

- The state of a qubit is described as a **vector**, which is a single-column matrix of size $n \times 1$ composed of complex numbers.

$$\begin{bmatrix} V1 & V2\cdots & Vn \end{bmatrix} \qquad \begin{bmatrix} V1 \\ V2 \\ \cdots \\ Vn \end{bmatrix}$$

Row vector   Column vector

- Vectors are multiplied together through the **inner product** that maps two vectors in a complex number or through the outer product that maps them to a linear transformation. It describes how to express one vector as a sum of simpler vectors.

- A **linear transformation** in one qubit is an operation in the qubit sphere that rotates vectors within it.

- **Eigenvalues** of a linear transformation are special complex numbers that provide a change in vectors when transforming them in quantum mechanics operations linking physical operations.

- **Eigenvectors** of a linear transformation are special non-zero vectors that maintain their directions after being transformed by the matrix of the linear transformation. They are used to encode, manipulate, and measure information.

- **Tensor products** are mathematical operations that combine two or more vectors or matrices. They are used in a quantum circuit to combine multiple qubits with quantum operators to transform them from one state to another and compute.

- Consider two vectors U and V, their tensor product, denoted U⊗V results in a matrix:

$$
U = \begin{bmatrix} a \\ b \end{bmatrix}, C = \begin{bmatrix} c \\ d \\ e \end{bmatrix} \rightarrow \begin{bmatrix} a \\ b \end{bmatrix} \otimes \begin{bmatrix} c \\ d \\ e \end{bmatrix} = \begin{bmatrix} ac \\ ad \\ ae \\ bc \\ bd \\ be \end{bmatrix}
$$

Quantum Computers Overview

A QUANTUM COMPUTER, LIKE A classical computer, has a set of techno-logical layers, each with a specific function to make the system work. We explain the different layers of a quantum computer, simplifying them in Figure 2.1, Figure 2.2, Figure 2.3, and Figure 2.4.

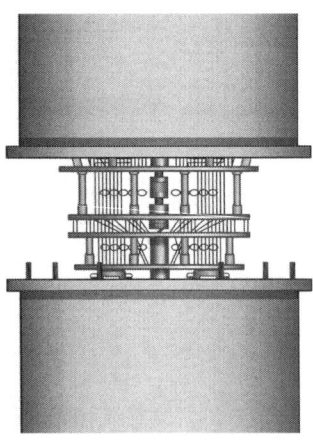

FIGURE 2.1   Outside shell cover.

Quantum computers are insulated from the external environment through shell covers that maintain cool conditions necessary for qubit operations. This forms a dilution refrigerator that protects the computers from outside disturbances and cools it down to millikelvin (mK) temperatures. Dilution refrigerators achieve their cooling power through the use of helium.

The refrigerator has gold-plated copper plates and features several stages, which achieve progressively colder temperatures from top to bottom. The lowest level stores the quantum chip, which must operate at very low temperatures to preserve the quantum information.

DOI: 10.1201/9781003302674-4

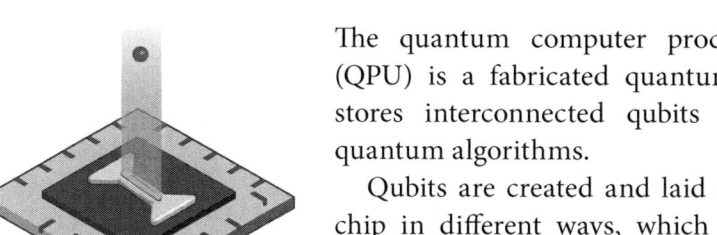

FIGURE 2.2 Quantum processing unit.

FIGURE 2.3 Physical qubits.

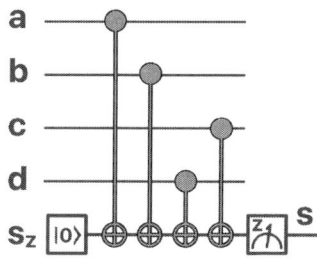

FIGURE 2.4 Quantum circuits.

The quantum computer processing unit (QPU) is a fabricated quantum chip that stores interconnected qubits to perform quantum algorithms.

Qubits are created and laid out into the chip in different ways, which affects how they operate depending on their connectivity, gate fidelity, and coherence time. QPUs require a specific infrastructure to operate, which suppresses external variables, and they are controlled by complex systems that must cross physical barriers to the processor.

Physical realization of qubits can be reduced to two-level energy state atoms, the ground state with lower energy, and the excited state with higher energy. To manipulate them, we change their spin states by exposing qubits to periodic electric or magnetic fields during specific time intervals. This is known as Rabi oscillations.

They are manufactured differently using either a real atom or an emulated one, and the biggest challenge is maintaining their coherence to prevent their states from dissipating.

Quantum circuits are visual representations of a wired diagram of gates that transport and manipulate information. The circuit size refers to the number of gates, its depth denotes the maximum length of a directed path from the input to the output, its width indicates the maximum number of gates that act in onetime step and its height represents the number of physical qubits involved in the computation.

A quantum gate is represented by a "box" where the qubit operation takes place to change its state, with a set of input and output lines corresponding to the number of qubits involved in that computation.

2.1 ANALOG AND DIGITAL QUANTUM COMPUTERS

There are two main approaches for programming quantum computers, (1) analog quantum computers and (2) digital quantum computers. The terms "analog" and "digital" refer to two different processing methods. The first one operates by creating a continuous representation of the physical world. The latter represents information using discrete numerical values (often in binary form) and perform calculations using digital logic. To visualize the concept, we provide an example of a clock in Figure 2.5.

FIGURE 2.5   Concept visualization of analog vs. digital concept.

The primary difference lies in how they display the time:

- In an analog clock, the movement of the hands continuously represent the passage of time, directly analogous to the movement of shadows on a sundial.

- In a digital clock, the change in numbers on the display represent the passage of time in discrete increments, reflecting a specific, quantifiable time unit passed.

Analog computers represent data using physical quantities, they perform operations using continuous functions. They were highly regarded for their capability to simulate and resolve issues in real-time for specific fields, such as aerospace for flight simulations.

Digital computers represent data using binary digits, perform operations using discrete steps. They are more versatile than analog computers, as they can perform any computation, provided it is given enough time and memory resources. This is what is known as universal computer, tied to Alan Turing's Turing machine concept, a theoretical model to have a machine capable of performing any possible mathematical computation.

Quantum computers can also be analog or digital. The main difference is that, the first one is a continuous method and the latter a gate-based sequence of discrete operations.

Analog quantum computers, such as quantum annealers, operate by gradually modifying the energy levels. They apply the quantum adiabatic theorem that states that a quantum system can remain in its lowest energy state, or ground state, if changes to the system are made slowly enough.

Quantum annealing addresses optimization by mapping problems onto the system's energy, aiming for the lowest energy state, which is the optimal solution. It leverages quantum tunneling, allowing the system to bypass local minima in search of the global minimum, as can be seen in Figure 2.6.

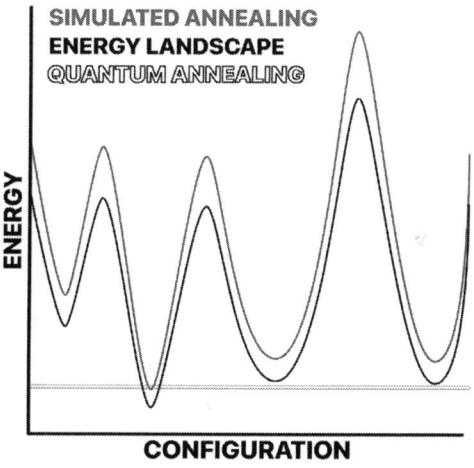

FIGURE 2.6   Visualization example of the annealing process.

Digital quantum computers, run algorithms through quantum gates and circuits, to manipulate qubits in a sequence of discrete operations. These gates change qubit states via energy excitation techniques, using pulses of electric or magnetic fields at precise frequencies. They require sophisticated equipment and fine-tuning to ensure that the quantum states are manipulated as intended. Quantum gates alter qubit states to perform operations in the same way that classical logic gates do, and they also have specific functions, as illustrated in Figure 2.7.

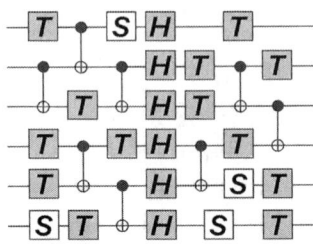

FIGURE 2.7   Visualization example of digital gates.

2.2 QUANTUM COMPUTER SIMULATORS

Richard Feynman pointed out that for a classical computer to simulate a quantum computer, it must explicitly represent each of the inputs that a quantum computer manipulates in a superposition of an exponential number of possible inputs.

A quantum simulator is a **software solution** that mimics the behavior of a quantum computer and runs on a classical computer instead of a quantum one. It is an SDK that allows to test different operations and provides an intuitive appreciation of how a real quantum computer can operate. By tweaking the parameters of the simulator, developers can explore different aspects of the system's behavior. Quantum simulators run on powerful HPC machines as they require high processing power and large amounts of memory.

At a high-level, a quantum simulator can be created in five basic steps described below.

1. Choose the computation to perform with the simulator.

2. Represent the computation as a circuit composed of quantum logic gates.

3. Use unitary matrices to simulate and describe how logic gates influence the operation of qubit states.

4. Determine the memory register size and initialize it.

5. Evolve the computer, test the computation, and extract the reading.

Quantum simulators are powerful tools used by developers as they have no noise and are helpful to estimate quantum algorithms. For example, they can be used for:

- **System modeling and interaction analysis:** simulating the behavior and interactions of qubits within a quantum system to enable understanding qubit entanglement, superposition, and decoherence dynamics.

- **Algorithm testing and optimization:** testing and refining quantum algorithms to evaluate algorithm performance, determining computational requirements.

- **Error simulation and correction:** simulating potential errors that affect quantum systems to test and assess various error correction schemes.

- **Resource estimation:** estimating the quantum computation resources by determining the number of qubits and their specific gate configuration.

- **Hamiltonian dynamics modeling:** modeling the dynamics of a quantum system under the effect of complex interactions.

The primary limitation of quantum simulator software is its scalability as running it in classical computers are limited by memory and processing power constraints. If simulations are approximations with non-fully entangled systems, or uses simpler gates such as Hadamard or C-NOTS, then the qubit limitation is not that relevant.

2.3 QUBIT MODALITIES DEFINITIONS

Qubits can be physical or logical. A **physical qubit** operates as a two-level quantum system and serves as a fundamental element in quantum computing. A **logical qubit**, or an abstract qubit, functions in accordance with the specifications of a quantum algorithm. It maintains coherence for a sufficient time, making it compatible with the operations of quantum logic gates. These are the qubits that are used to perform operations.

Qubit modalities are typically referred to as computing platforms, which are the different ways to create a qubit. To create a qubit, scientists need to gain control and manipulate the electron spins with different methods, such as the microwaves or magnetic fields. They need to be easily controlled from the outside, to perform calculations, and enable the extraction of computing results.

When performing operations, in addition to the number of qubits needed, it is important to consider that qubits need to be isolated so that they do not interfere with each other and can maintain their properties. They need a certain quality for rotations, the entanglement efficiency (a combination of lifetime and accuracy), the time needed for the qubits to remain in a useful state for operations, and the gate speed needed to operate before the system decoheres.

This quality concept, known as **qubit performance**, is described by five criteria established by David DiVincenzo in 2000, and they are considered as prerequisites for evaluating physical systems that may be used to realize quantum computing.

1. Using well-characterized and scalable qubits that can operate properly.

2. Initializing qubits to consistently prepare for the same state with minimal error.

3. Maintaining the qubit's properties long enough to perform operations.

4. Applying a universal set of gates to perform arbitrary operations on the qubits.

5. Ensuring the algorithm can properly read the results of the chosen qubits.

Understanding the performance of each qubit modality is important when coding an algorithm, as quantum gate performance is specific to qubit technology platforms, and some algorithms need fault-tolerant quantum computers to work properly.

Qubit modalities are at different maturity stages; some are still being researched to find their potential suitability, others are being tested to validate that they can work, and others are running in a commercial environment. They are described in Figure 2.8.

Commercial Phase

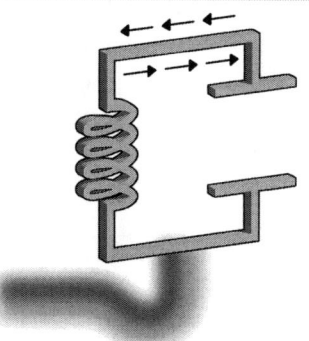

Superconducting qubits mimic atoms using special materials such as aluminum or niobium that conduct electricity without resistance.

They are manipulated by cooling to mK temperatures and using electromagnetic pulses to control them.

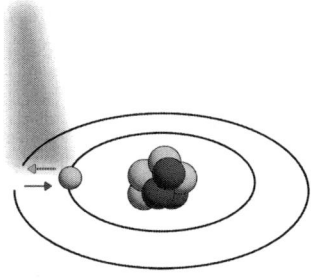

Trapped ion qubits are real atoms from the periodic table, specifically, chemical elements. They are trapped in an electromagnetic field using static voltages.

They are manipulated by cooling and are kept in a very low-pressure environment with high vacuum conditions and lasers are used to excite them.

Test Phase

Photonic qubits are individual particles of light (photons), that encode quantum information in a ring along with the scattering unit.

Lasers are used to manipulate the states through a phenomenon called "quantum teleportation" utilizing beam splitters.

FIGURE 2.8   Qubit modalities.   (*Continued*)

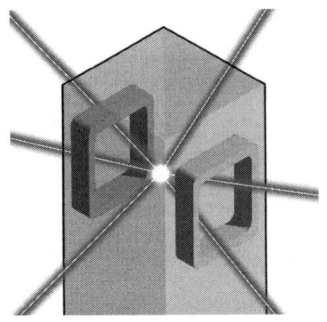

Neutral atom qubits are one-atom states (single ultracold neutral atoms), such as rubidium, arranged in configurable arrays.

They are manipulated by light beams to encode and read out quantum states, trapped with controlled laser beams and magnetic fields.

Research Phase

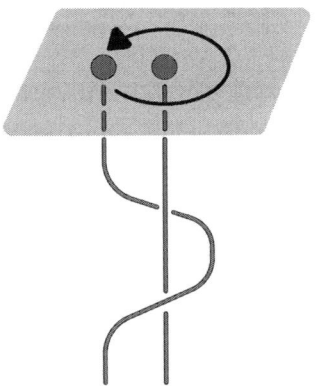

Topological qubits are created using Majorana fermions anyons, a type of quasiparticle. They are connected with each other by their topology.

They are manipulated by braiding anyons to form logic gates that transform the quantum state of the system and store information.

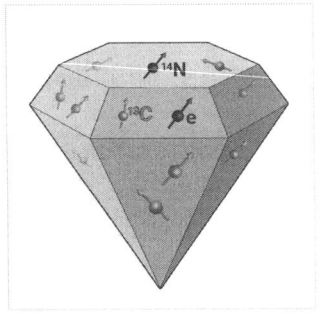

Diamond qubits are created by introducing a single nitrogen atom and a neighboring vacancy defect into a diamond lattice structure.

They are manipulated by applying microwave and laser fields to change the electron spin at its center.

FIGURE 2.8 (*Continued*)   Qubit modalities.

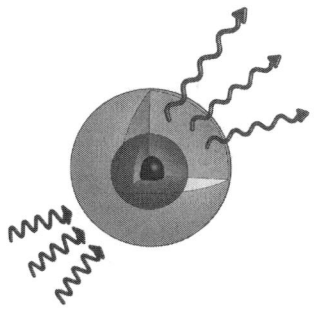

Quantum dots are created by confining electrons in a semiconductor in which they move through the material, conducting electricity.

They are cooled to liquid helium temperatures and optically excited. Their spin states are used as information stores, and laser pulses are used to read and alter the states.

FIGURE 2.8 (*Continued*)   Qubit modalities.

Technology providers are working on different qubit modalities with the hope of finding the one that will scale properly. Table 2.1 shows a (non-exhaustive) list of providers and their different qubit modalities as of the date this book was written.

TABLE 2.1 Companies Qubit Modalities (Non-exhaustive)	
Commercial Phase	
Superconducting qubits	Alice & Bob, Anyon Technologies AWS, Baidu, D-Wave* IBM, IQM, Google, Origin Quantum, Oxford Quantum Circuits, qci, Qilimanjaro*, QuTech, QuantWare, Rigetti, Riken, SeeQC, Toshiba
Trapped-ion qubits	Alpine Quantum Technologies, eleQtron*, Infineon Technologies, IonQ, Quantinuum, Oxford Ionics, Universal Quantum
Test Phase	
Photonic qubits	Orca, Psi Quantum, Quandela, Qui Quantum, Quantum Source, Quix, Xanadu
Neutral-atom qubits	Atom Computing, Infleqtion, Pasqal*, planqc, QuEra*
Research Phase	
Topological qubits	Microsoft, Nokia

(*Continued*)

TABLE 2.1 (*Continued*) Companies Qubit Modalities (Non-exhaustive)	
Diamond qubits/ NV	Quantum Brilliance, SaxonQ, XeedQ
Quantum-dot/spin qubits	Diraq, Equal1 Laboratories, Intel, Quantum Motion, SemiQon, Silicon Quantum Computing
*Note that these companies are building analog quantum computers rather than digital ones.	

In addition to performance, qubits must be able to scale well, enabling a large quantum computer. This is strongly dictated by the **qubit topology**, which we will explain in next section 2.4 "Different Qubit Topology". The choice of topology is influenced by the physical implementation of qubits and the intended computational tasks, aiming to maximize coherence times, minimize error rates, and enhance scalability.

2.4 DIFFERENT QUBIT TOPOLOGY

Qubit topology is the general layout of the qubits; it refers to the arrangement and connectivity of qubits within a quantum computing system. It dictates how many neighboring qubits a given qubit can interact with. If two qubits cannot directly interact with each other, we can use "swap" gates to enable virtual connections, but this approach adds overhead that can increase the error rate and limit scaling.

Topology structure is crucial because it determines how qubits interact and how quantum information is processed and transferred. It influences quantum computers architecture, algorithms execution, and error correction codes; this is why, apart from the qubit modalities, providers also bet on qubit arrangements (topologies).

Different topologies offer various advantages and trade-offs in terms of connectivity, error rates, and quantum computing performance:

1. **Straight line:** each qubit interacts only with its immediate neighbor. It is a simple configuration and easy to build, but it limits the operation efficiency if qubits are not adjacent.

2. **Two-dimensional lattice:** each qubit might be connected to four neighbors in a square grid. The layout is versatile since it allows more complex qubit interactions.

3. **Hierarchical tree structure:** It features a central root qubit that branches out to other qubits in a "parent–child" configuration. It works well with hierarchical style algorithms that require structure.

4. **All-to-all:** each qubit is connected to every other qubit, offering flexibility for quantum operations. This is ideal for complex quantum algorithms that require extensive qubit interactions, but it's challenging to implement due to the rapidly increasing number of connections with each added qubit.

5. **Ring:** all qubits are arranged in a circle, with each qubit connected to two neighbors in a closed loop. This is a simple structure that offers more connectivity.

6. **Modular:** all qubits are grouped into clusters with dense internal connections and fewer connections between the clusters. This helps with scalability and enables modular computation where different clusters handle different algorithm parts

The topology design depends on the specific requirements of the quantum algorithm, the physical implementation of qubits, and the desired balance between connectivity, error rates, and scalability, with no one-size-fits-all approach.

CHAPTER 3

Quantum Programming

Q UANTUM COMPUTERS ENCODE INFORMATION in a different way than classical computers. Qubits represent the smallest unit of quantum information containing one bit of classical information accessible by measurement.

For example, on a classical computer, three bits represent eight different state combinations, which are 000, 001, 010, 100, 011, 101, 110, and 111, but only one at a time. When we process information, we pick one of the states as the input, process it, and obtain result as the output. If we also want to process information using a different input state, we can process it either in parallel using additional hardware or sequentially using additional time.

On a quantum computer, three qubits can represent aspects of all eight different components in a single quantum superposition state; therefore, we have a quantum version of parallelism, as defined by David Deutsch. This scenario enables us to conduct 2^N calculations simultaneously with just an N-qubit processor instead of needing N-classical processors running simultaneously to perform the same number of calculations.

Figure 3.1 illustrates how a quantum computer calculates in "parallel".

DOI: 10.1201/9781003302674-5

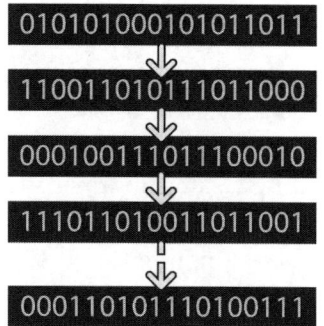

Classical computers iterate sequentially over all the values to find the right one.

Quantum computers can look up all values at once to find the right one in high speed.

FIGURE 3.1   Processing quantum vs. classical.

3.1 CIRCUITS DEFINITION

Quantum algorithms are represented as quantum circuits, a concept defined more than two decades ago. Circuits are used to describe quantum algorithms, and they help to quantify the number of gates and type of gates required. Their basic structure comprises qubits, gates, and measurements, as shown in Figure 3.2. The horizontal lines represent qubits, where the gate operations occur. Ordered from left to right, the bits of classical information are stored in classical bits represented by the double lines.

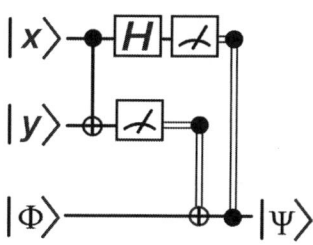

Initially, qubits are prepared and loaded in the states |x>, |y>, where |> represents the Ket notation.

Then, quantum gates are applied to manipulate qubit states and perform operations.

Finally, an output measurement is performed to find the circuit outcome. After the measurement, information is stored in ancilla (auxiliary) classical bits.

FIGURE 3.2   Quantum circuit structure.

In the above general quantum circuit structure, since we have two qubits, we have two different outcomes with probabilities between 0 and 1.

As we amplify the values, the one that is closest to 100% corresponds to the qubit that holds the state giving the answer to our problem.

We use interference to control the phase of each qubit and amplify the probability distribution so that it can reach a value close to unity, which corresponds to the state that gives us the answer to our problem. Each qubit outcome measures the probability of finding a qubit in any of the two-basis states, 0 or 1.

When the qubit values are read, the values are amplified, and their phase is modified to reinforce or cancel out the values, which mathematically helps to find the desired solution.

3.2 GATES FUNCTIONING

Quantum gates are operations that transform the state of qubits. When we operate on a qubit to rotate the state vector within the Bloch sphere, to encode information and perform operations, the spins of the electrons are hung around the X, Y, and Z axis at different angles. The rotation categories are shown in Figure 3.3.

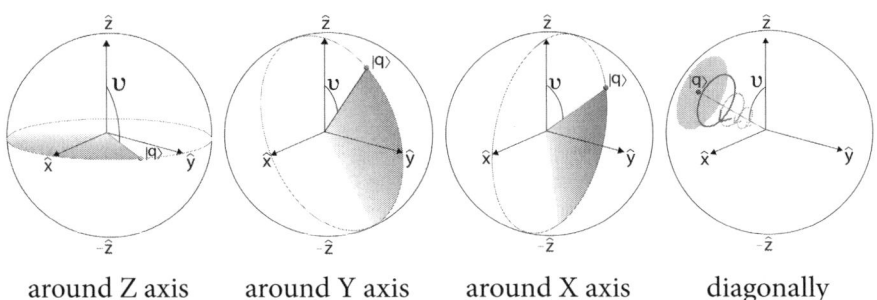

    around Z axis   around Y axis   around X axis   diagonally

FIGURE 3.3   Qubit rotations.

Programming gates are devices that perform basic logical functions that are fundamental for digital circuits. They make decisions based on a combination of digital signals coming from their inputs. Depending on the type of logic gate being used and the combination of inputs, the binary output will differ. Any quantum operation can be written approximately using a universal set of gates, which in principle can implement any type of quantum algorithm.

These gates manipulate information on qubits and are the basic components of quantum algorithm design. They are operators that act on the states of a certain qubit.

A quantum gate is represented by a "box" where the operation occurs, with a set of input and output lines that equal the number of qubits involved in that computation.

Gates can act over 1 or many qubits. Specifically, we have:

- **Single-Qubit Gates:** 1-input 1-output. One-qubit gates take one input qubit and transform it into one output.

- **Two-Qubit Gates:** 2-input 2-output. Two-qubit gates take two qubit inputs and entangle them to perform operations.

- **Three- or more-Qubit Gates:** N-input N-outputs are combinations of controlled and target gates that operate on multiple qubits.

Table 3.1 shows some of the most important qubit gates, illustrated using Quirk open-source drag-and-drop circuit editing tool https://algassert.com/quirk.

Toffoli and Fredkin are universal reversible gates, this means that they have number of inputs and outputs, which minimizes energy loss during computations. It is a very versatile gate that can be used to simulate other gates. On the other hand, H, T, S, and control NOT form a universal set of gates that are independent of one another. By combining them, we can derive the remaining gates. This enables the quantum computer to execute any quantum operation.

Interestingly, quantum gates do not erase information during computation, because, they are **reversible**. We have a unique input associated with a unique output, and vice versa. Based on a gate's output, we can infer its input. This means that a computation based on reversible logic gates can be run forward to obtain an answer, the answer can be copied, and then we can recover all the energy used. The fact of **requiring less energy to run** is a great advantage over classical computers. A classical computer takes 97 MWh to run, while a quantum computer takes only 0.00042 MWh to run. This is explained by Landauer's principle, which states that dissipation of energy is necessary only when information is erased. This is what classical computers do for each logical operation. The energy usage of the quantum system does not scale directly with the number of qubits. Even if it grows to 10,000 qubits, it is estimated that would require only 100 KW.

TABLE 3.1   Main Quantum Programming Gates

1 qubit	Pauli gates (X, Y, Z)	Rotates around the X, Y, and Z axis. The X gate is called a *bit-flip* NOT gate and it connects gates to perform superposition.	
	H Hadamard	Achieves a qubit superposition used to load data by rotating the qubit in a diagonal plane across the X–Z axis.	
	Z, S, T phases	Shifts the gates, rotating around the Z axis.	
+ 1 qubit	Controlled Pauli: control Y, control Z, control X	Achieves qubit entanglement by applying a Y, Z gate or X gate (NOT) to the target qubit, if the control qubit is in state 1.	

(*Continued*)

TABLE 3.1 (*Continued*) Main Quantum Programming Gates			
	SWAP	It exchanges (swaps) the states of two qubits.	
	Toffoli or controlled-controlled-NOT gate (CCNOT)	It is a logic gate that is reversible, i.e. with no information loss. If the first two bits are both set to 1, it inverts the third bit; otherwise, all bits stay the same.	
	Fredkin or controlled SWAP (CSWAP)	It swaps the values of the second and third qubits if the first qubit is in state 1.	

3.3 CIRCUIT REQUIREMENTS

To understand the performance of the algorithms, we must analyze how many gates it has, since they will need more pulse control to change their states, and maintaining quantum coherence is difficult. Quantum errors can manifest as bit or phase flips and a combination of both during the algorithm execution. Quantum algorithms are made by quantum circuits, which in turn face many challenges. These includes both in terms of the size of the calculable problem related to the number of qubits, and the extent of simulation related to the coherence time and accumulation of errors.

Circuit requirements must be analyzed in terms of depth and length.

- Circuit length refers to the total number of quantum gates in the circuit, sequential or parallel, and measures the overall complexity of the circuit.

- Circuit depth is a measure of the longest path in the circuit, representing the total number of gates in sequence on the qubits. It affects the error rate due to decoherence, which will be described later in Section 4.6 'Algorithm Noise' of this book.

In the circuit shown in Figure 3.4, the depth is 4, which is the longest sequence of calculations, while the length is 7, which is the total number of gates in the circuit.

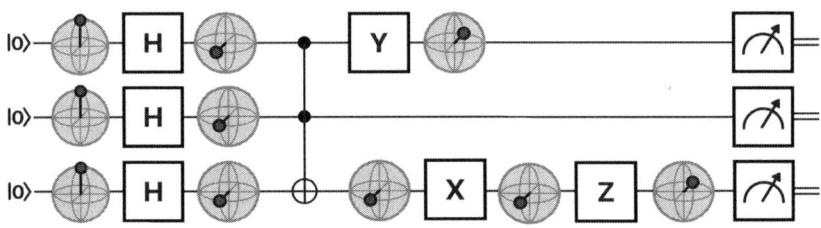

FIGURE 3.4   Circuit depth and length.

The depth contributes to the complexity analysis of the circuit, the performance since they take longer to execute, contributes to circuit size which affects computational requirements.

Another aspect to consider is the **complexity of the gates** since some join several qubits and have stronger hardware requirements. For instance, the Toffoli and Fredkin gates, which apply to three qubits may be split into a collection of "subgates".

An important item to consider is **uploading data** into the quantum circuits, which must be done by encoding it into qubits, a non-trivial task that balances the number of required qubits and the circuit runtime complexity. Both the data itself and the encoding influence the runtime of the loading process; in the worst case, loading requires exponential time, which influences the algorithm performance and speedup. There are many techniques used to encode data to upload it to the quantum circuit. It can be mapped into qubits or their amplitudes, or gates, either in superposition or rotation.

At the end of a quantum circuit, data is read through a process called quantum measurement or qubit readout. This process collapses the quantum state of a qubit into a classical state, translating quantum information into classical information that we can understand and use.

3.4 PROGRAMMING TOOLS

Quantum software development kits (SDKs) provide collections of tools to develop quantum circuits, apply quantum gates, and measure qubits. They also provide libraries, frameworks, and integrations with classical languages and platforms. They give developers the ability to build, compile, run, and analyze quantum circuits and quantum programs.

These SDKs have five common characteristics:

1. Most are written in Python, making them accessible to a wider community given the programming language popularity.

2. They enable the creation and manipulation of quantum circuits.

3. They offer simulations on local machines or the cloud, and some can connect to hardware.

4. Many of them are open source, fostering community collaboration and innovation.

5. They provide tools for specific algorithms such as optimization and machine learning.

Some of them are developed by startups, academic institutions, or other technology manufacturers.

Table 3.2 shows a non-exhaustive key development platform as of 2023.

TABLE 3.2 Quantum Computing SDK List (Non-exhaustive)	
Name	**Creator**
Amazon Braket Python SDK	Amazon Corporation
Cirq	Google Corporation
cuQuantum	NVIDIA
Forest	Rigetti Startup

(Continued)

TABLE 3.2 (*Continued*)   Quantum Computing SDK List (Non-exhaustive)

Name	Creator
Intel Quantum SDK	Intel Corporation
Ocean	D-Wave Startup
Orquestra	Zapata Startup
PennyLane	Xanadu Startup
Perceval	Quandela Startup
ProjectQ	Institute for Theoretical Physics at ETH Zurich
Q# Azure Quantum Development Kit	Microsoft Corporation
Qibo	Several Research Laboratories
Qiskit	IBM Corporation
Strawberry Fields	Xanadu Startup
t|ket>	Quantinuum (merge between CQC Startup and Honeywell)

As technology evolves, new layers of abstraction appear that make it easier to access computing systems, program and run quantum circuits. There exist cloud computing containers that execute packages in an application's code along with all its libraries and dependencies, to allow easily running that application on any system. These cloud systems allow pay-as-you-go easy access to quantum systems and simulators, with ready-to-use available code.

CHAPTER 4

Quantum Algorithms Overview

S IMILARLY TO CLASSICAL SYSTEMS, quantum algorithms are a set of step-by-step instructions or rules designed to perform a specific task or solve a particular problem. They process information to solve a problem, by encoding information in the state of a single qubit, or the joint state of many qubits, and performing operations over them. Some algorithms are simple instructions and enable other algorithms, while others consist of many instructions to address more complex problems. To properly understand how algorithms work, it is important to take some factors into account, shown in Figure 4.1, and which we review in this section.

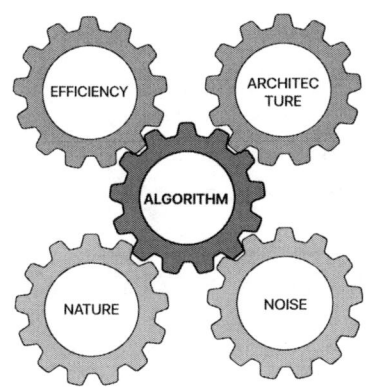

FIGURE 4.1   Algorithm factors.

DOI: 10.1201/9781003302674-6

- **Efficiency** is related to the number of computational resources required by the algorithm; it could be related to the speed or resources needed.

- **Architecture** is related to the algorithm design, which specifies its different parts and where it runs, in an HPC or quantum computer.

- **Noise** is a persistent problem that affects algorithm performance, creating poor convergence or suboptimal solutions.

- **Nature** of the algorithm refers to whether it is a quantum or quantum-inspired algorithm, which makes use of quantum mechanics but runs in a non-universal quantum computer.

Like classical computers, which use a variety of algorithms tailored to specific tasks, quantum computing also employs diverse algorithms designed for different purposes, which is described in Section 3.3 of this book.

4.1 HYBRID APPROACH

Quantum computing hybrid approaches refer to the concept of a quantum computer and a classical computer working together to solve a problem in a back-and-forth mode, as seen in Figure 4.2.

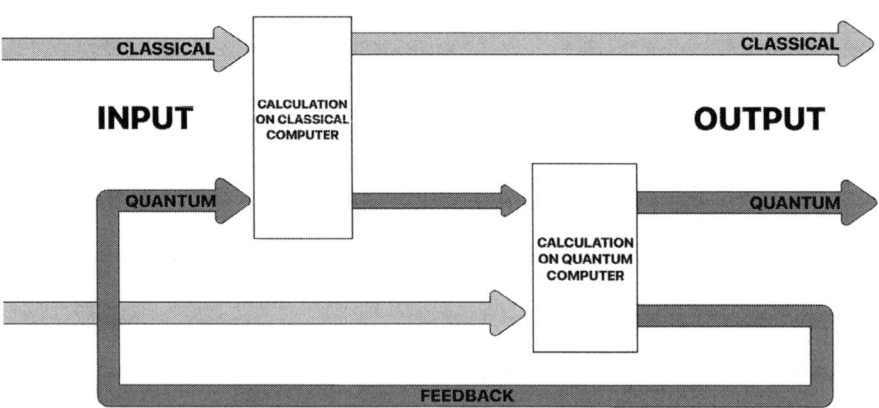

FIGURE 4.2   Conceptual visualization of hybrid approaches.

The concept is understood and defined in many ways. For clarity, this book simplifies it into two definitions, one focused on software and the other on hardware.

4.2 SOFTWARE APPROACH

It refers to the variational quantum algorithms (VQAs), a class of quantum algorithms that leverage both classical and quantum computing resources to find approximate solutions to problems. They leverage classical computers (HPC) to accelerate some of their subroutines.

Common types of hybrid algorithms include the Variational Quantum Eigensolver (VQE), the Quantum Approximate Optimization Algorithm (QAOA), and the Quantum Neural Network (QNN), which are explained later.

These algorithms use a quantum computer to prepare states and assess their properties, while a classical optimizer adjusts the parameters of the quantum operations to find an optimal solution.

4.3 HARDWARE APPROACH

It refers to how both hardware types, HPC and Quantum Processing Unit (QPU) resources work together to run hybrid algorithms through a middleware abstraction layer. This layer manages the whole computation process to enable seamless, unified resource allocation and management. The different layers are shown in Table 4.1.

TABLE 4.1   Hybrid Computing Layers
Host HPC: interfaces with the user and stores data
Processor: performs the algorithm iterations.
Measurement: controls all operations results.
Quantum: stores the qubits for computing.

It is essential to have a comprehensive hardware that provides fast access for quantum control and leverages HPC for Quantum Error Correction (QEC) at scale. This integration is done using an API to communicate with the CPU of the HPC system.

There are different aspects to consider when building the middleware:

- **Integration:** Real-time integration to exchange data for processing an algorithm, which is challenging as there are some inherent limitations to the cloud access to quantum computers.

- **Acceleration:** The use of the QPU kernels to accelerate a specific type of task, which must run in parallel feeding into each other, like the case of quantum circuits such as VQAs.

- **Automation:** Splits the workflow of tasks between both systems in which the QPU kernel is integrated without modifying the HPC application, as part of an end-to-end computation.

In this setup, a Quantum Processing Unit (QPU) is tasked solely with running quantum circuits while other key tasks remain in the realm of classical computing: translating problems into quantum-compatible formats, managing the program translation (via a transpiler), correcting errors, and interpreting the QPU's outputs for practical use.

4.4 QUANTUM INSPIRED ALGORITHMS

Quantum-inspired algorithms are based on quantum physics and run on classical hardware. They are commonly used in mathematics, computer science, condensed matter physics, and now in quantum information science. Some of the problems solved are related to optimization, machine learning, and molecular analysis.

Quantum-inspired algorithms are a good way to start implementing quantum solutions that can later be migrated to a quantum algorithm. They can do the following:

1. Emulate quantum algorithms that cannot run on quantum computers due to hardware limitations.

2. Provide insights into quantum algorithms requirements to help in developing more efficient algorithms.

3. Improve quantum algorithms performance verifying the computation fidelity on hardware devices.

4. Analyze if there is a more efficient quantum-inspired solution than the potential quantum algorithm advantage.

Tensor networks are an example of commonly used quantum-inspired algorithms to simulate quantum systems. They are mathematical functions of multiple parameters that are linear to each other and use multidimension matrices to manipulate complex multidimensional data to replicate entanglement and superposition properties. They are good for mimicking quantum systems since they operate similarly and even have a notation diagram that connects components with wires just like quantum circuits.

Figure 4.3 shows a visual example of the similarities in notation.

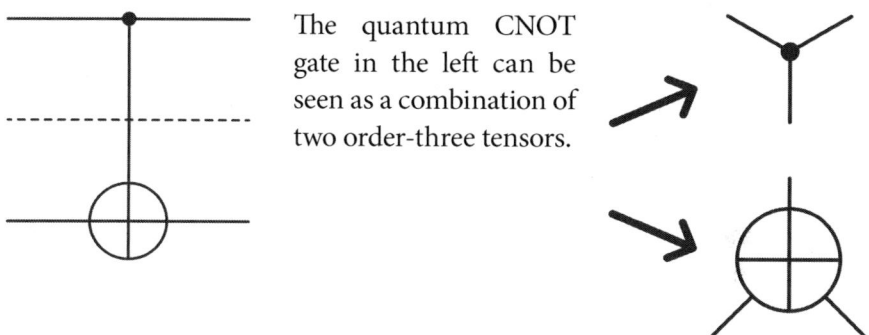

The quantum CNOT gate in the left can be seen as a combination of two order-three tensors.

FIGURE 4.3   Notation: tensor network and quantum circuits.

Another related quantum-inspired concept is a digital annealer classical computer that simulates the behavior of an annealer, which in essence is a hardware machine rather than an algorithm, and which is commonly used.

In both algorithm and hardware cases, quantum-inspired algorithms cannot fully replicate the power of quantum computers since they are limited by the nature of classical systems, but they can offer some improvements on certain classical algorithms.

4.5 ALGORITHM PERFORMANCE

The goal of quantum algorithms is to create programs that **outperform** their classical counterparts so that they can show a quantum advantage, demonstrating a performance improvement over classical computers for certain problems in terms of speedup or efficiency.

To assess whether an algorithm outperforms another algorithm, we must first understand the concept of **problem complexity**, which mathematically quantifies the computational cost; it is measured by the running time to solve the problem and is related to the number of steps required to solve the problem (time) and the space the process takes (memory). Moreover, another complexity factor considered is the energy needed to perform a computation, which scales with the problem size. Typically, as the problem size grows, so does the number of steps needed to solve it. This means that it takes more time and may become infeasible.

Problem complexity can be divided into two categories, polynomial and exponential, as defined in Figure 4.4. The main difference is how the number of solving steps increases with the problem's input size.

Polynomial problems → Easy-to-solve problems

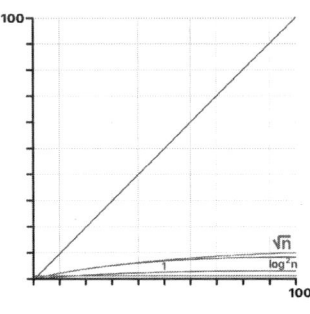

Problems that grow more slowly than the problem input size. The number of steps to solve the problem is, in the worst case, a polynomial in the size of the input.

Examples:

N linear, e.g., search an unsorted list $\sqrt{n}$ square root, e.g., matching **Log(n)** logarithmic, e.g., binary search **C** constant, e.g., fixed interest rate

Exponential problems → Hard-to-solve problems

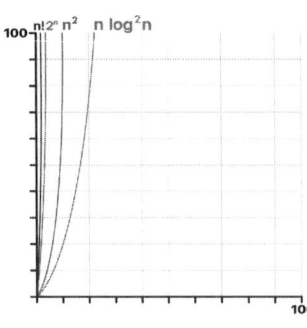

Problems that grow faster as the size of the input increases. The number of steps to solve the problem is a multiplication, with the best case being the square root of the input.

Examples:

n! factoring, e.g., city routing 2^n size doubling, e.g., decryption n^2 quadratic growth, e.g., checking duplicates **n Log(n)** log-linear, e.g., sorting in a specific order

FIGURE 4.4   Problem complexity solving time.

Reducing the algorithm complexity to reduce its growth rate would result in a **quantum advantage**, when the quantum computer can perform a particular computation considerably faster than even the best classical computer. When the quantum computer can perform computations which no classical computer can perform at all, then we run into what is known as **quantum supremacy**. It implies that a quantum computer can complete a specific task or computation much more rapidly than any classical computer, rendering classical computers virtually unable to compete.

Google demonstrated the quantum supremacy concept in 2019 with their quantum processor, Sycamore, a 53-qubit chip. The experiment involved sampling the output of a random quantum circuit, a method used to study properties of quantum systems and assess the performance of quantum hardware, which was chosen as it is computationally hard for classical computers to simulate efficiently because it would take an approximately 10,000 years for an ultra-fast super-computer.

Simulating such a random quantum circuit classically would require an exponential amount of time as the size of the circuit increases. Google's Sycamore quantum processor was able to complete this task in a polynomial amount of time, specifically in about 200 seconds.

Quantum computing can be hundreds to thousands of times quicker than classical computing, require only a minimal portion of the memory needed by a conventional computer, or execute tasks that are currently unachievable with today's technology.

- Speed implies that it takes less time to perform a calculation by reducing the problem growth, either in an exponential or polynomial way.

- Memory storage is increased since less space is needed to store information by carrying it with fewer qubits, specifically about $\log[N]$.

- Unachievable means the resolution of a problem cannot normally be tackled or solved today with a classical computer due to the problem size.

When describing the theoretical improvement of a quantum algorithm, we compare mathematically the algorithm complexity reduction and speedup improvement. Figure 4.5 shows a visual representation of algorithm complexity reduction.

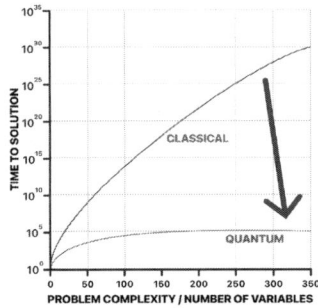

Exponential Speedup:

The quantum algorithm reduces the complexity to a polynomial growth from an exponential one, flattening the curve.

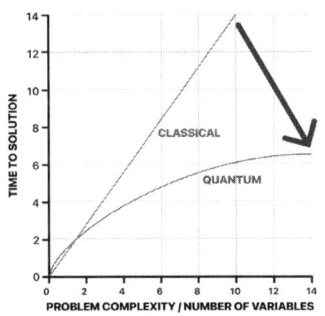

Polynomial Speedup:

The quantum algorithm reduces the polynomial complexity to a slower growing one.

FIGURE 4.5   Quantum advantage.

4.6 ALGORITHM NOISE

In technology, noise disturbance interferes with data transmission and corrupts quality, creating errors. This problem is typically addressed using redundancy techniques that enable operationality to be maintained. Errors can occur due to various types of noise that can interfere with information during transmission, storage, or processing, ultimately leading to corruption. The nature and impact of this noise typically depend on the physical medium involved and the specific process being undertaken, i.e. software and mechanical malfunction.

A simple reliability problem can introduce errors into a system. In the illustrative example in Table 4.2, we show the impact of a simple bit-flip in the encoding of a letter that affects the whole word. By flipping a 0 by a 1 the letter turns from C to G, impacting many words that can be constructed with it, like for example Core versus Gore.

TABLE 4.2	Bit Error Impact Illustration
Letter C	Letter G
01000**0**11	01000**1**11
Core	**G**ore

The biggest challenge is how to deal with noise and correct errors without imposing a huge computational overhead that will eat up all the computing power rather than running the algorithms. In digital computers, this problem has been overcome and are under control.

For quantum computers, the challenge is to keep the qubit **coherence time**, which describes how long a superposition between two quantum states |0> and |1>. The problem is that the coherence time may not be long enough to run the operations in a quantum algorithm, so the information is lost. To estimate error rates, we must know the number of qubits and gates to operate, as shown below.

$$\text{Error rate} = \frac{1}{\#\text{ qubits} \times \text{circuit depth}}$$

As of today, when this book is written, quantum computers **errors** are typically around 10^{-3} (that is 0.01%) instead of 10^{-18} of classical computers. So, developers must be in the lookout for errors to correct them.

The engineering goal in quantum computing is to find good quality qubits that are almost noiseless.

Quantum noise is literally a measured vibration that can lead to inaccuracy. This means that the computations may not necessarily correct, and this propagates through all computations. Even small noise accumulates during computation, and quantum properties decay over time (decoherence), increasing noise levels.

Current **noisy intermediate-scale quantum (NISQ)** systems are limited by noise, which affects how we control those qubits and limits the size of quantum circuits that can be executed reliably. The limitations are the number of qubits N, the number of steps that can be executed, and the circuit depth d, described in Section 1.2 of this book. The term NISQ was famously coined by John Preskill.

Over time, we believe that by using multiple circuits working together to detect and correct errors, we can achieve a higher level of computation.

The end goal is to **construct a fault-tolerant quantum computer (FTQC)** that works effectively even when its elementary components are imperfect or even if some parts fail because they have small and constant gate errors that do not generate significant computing overhead.

Quantum error correction (QEC) is essential for reducing the effects of noise on stored quantum information, which can occur in faulty quantum gates, faulty quantum preparation, or faulty measurements. Several mitigation techniques are currently being tested, with the end goal to reduce overhead.

To **suppress errors**, information is encoded in a single qubit and distributed across other supporting qubits, forming a group of physical qubits referred to as "**logical qubits**", where the qubits work together to perform computations with a lower overall error rate. This procedure protects the integrity of the original quantum information and stabilizes quantum computers. Because individual qubits are fragile, one error reduction method is to entangle qubits to form one "logical qubit", which can be combined with other logical qubits into a fault-tolerant quantum circuit, as shown in Figure 4.6.

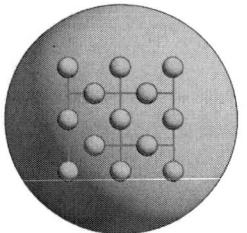

One logical qubit is made from several qubits that are entangled and applying several error correction techniques on them. This new logical qubit is then used in quantum circuits acting as a single qubit that can be combined with other logical qubits creating error free circuits.

FIGURE 4.6   Logical qubit visualization.

An alternative method to reduce errors involve **error mitigation** to minimize their impact by adjusting the measurement outcomes from a quantum circuit. Ultimately, the goal is to achieve an error correction similarly to classical computing that encodes information with redundancy to compensate when errors happen. A scheme to help mitigate errors is to couple qubits to another qubit that does not take part in the calculation called "ancilla".

Figure 4.7 shows the evolution of the error correction techniques, from smaller to larger improvement rates.

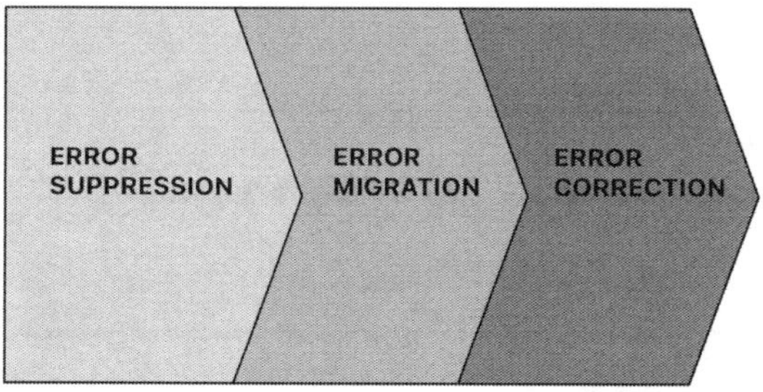

FIGURE 4.7   Evolution towards correction.

- **Error Suppression:** To reduce the likelihood of errors, strategies may involve selecting physical systems that are inherently more resistant to certain errors, or utilizing techniques like dynamical decoupling. Dynamical decoupling involves a series of operations designed to neutralize the impact of unwanted environmental interactions by effectively averaging them out.

- **Error Mitigation:** To mitigate the effect of errors on the end results, the strategy involves altering the quantum algorithm's parameters to produce various scenarios in which errors influence outcomes differently. These varied outcomes are then aggregated in a manner that diminishes the overall error in the final result.

- **Error Correction:** When an error arises, the system utilizes the redundancy embedded in the quantum code to detect and rectify it.

This process does not require knowledge of each qubit's state, thereby safeguarding the quantum data. However, this method necessitates the use of a larger number of qubits and entails more intricate operations.

The **threshold theorem** is a framework used to quantify the fault tolerance level to effectively perform quantum computations. As we correct errors, new ones appear, so we must correct them faster and eventually reach a critical mass that enables the process to be self-sustaining. It shows that a NISQ computer) can perform computations effectively, provided that the noise strength remains below a quantum accuracy threshold. According to quantum information theorist Scott Aaronson, computers are not currently self-sustaining.

Providers have developed techniques for **error detection** and for **quantum error mitigation**. This process is performed via classical software, but it can only be performed for small problems since the cost is exponential. Thus, this approach represents an interim solution for correcting errors, and the end-goal is to have an error-free quantum computer.

CHAPTER 5

Algorithm Foundations

Q UANTUM ALGORITHMS FOLLOW A sequence of instructions designed to perform a calculation or solve a problem using the principles of quantum mechanics described earlier in Section 1.2 of this book. Unlike classical algorithms, which operate on bits in a deterministic manner, quantum algorithms work on qubits that are influenced by probabilistic outcomes of quantum states. We could say that quantum computing algorithms are non-deterministic, in the sense that at any point where there are multiple possible actions, the non-deterministic machine can "choose" any of them, potentially exploring many different paths of execution simultaneously.

This is what is known as parallelism. Quantum algorithms use parallelism to encode data in quantum superposition, which enables them to "simultaneously" perform operations. They can find solutions through probability amplitudes. This process enables quantum algorithms to solve problems faster than classical computers as they may require fewer instruction cycles to find the solution.

Besides quantum gates to code algorithms, there is also a set of "pre-made" smaller algorithms considered functions that perform a specific action for problem-solving techniques. These are the **building blocks** of many algorithms and are useful for understanding how to manipulate data, perform calculations, and solve various computational problems.

On top of this, there is a set of fundamental algorithms that serve as the basis for more complex algorithms, because their underlying principles can be applied to a variety of problems.

DOI: 10.1201/9781003302674-7

These algorithms have shown the potential of quantum computing and they serve as :

1. A benchmark of capabilities to show the algorithm's quantum advantage potential

2. A foundation for subsequent research and development over the algorithm's operations

- **The unitary operator** (U_f) is a bounded linear operator, that preserves the inner product. It efficiently encapsulates multiple gates into one notation and provides a way to simplify a complex operation by "consolidating it". The operator is used pervasively across multiple quantum algorithms.

  It is the smallest unit which is used pervasively over all the algorithms.

- **The Quantum Fourier Transform (QFT)** is used to construct a linear transformation map of quantum states and amplitudes to perform quantum operations. It is used in quantum phase estimation, shor, linear systems, and many other algorithms.

  It enables an exponential speedup.

- **Amplitude Amplification (AA)** is applied to identify a target value encoded in a qubit by increasing its state value thus boosting the probability to close to 100%. This approach is the basis for many subroutines and algorithms, and it is notably used as the foundation for Grover's algorithm and many others.

- **Quantum Random Number Generator (QRNG)** it generates random numbers by using the principles of quantum mechanics, exploiting its inherent unpredictability of quantum phenomena and the fundamentally probabilistic quantum nature.

  It ensures true randomness.

- **Quantum Phase Estimation (QPE)** it estimates the value of the phase of a quantum state, with a probability that depends on the number of gates and qubits used. It determines the eigenvalues of a unitary matrix, which is needed to perform matrix calculations. This approach can be used in many applications, such as calculating electronic structures and variational quantum eigensolvers.

- **Quantum Monte Carlo** it simulates the quantum many-body effects that govern the motion of particles through the direct treatment of their electronic wavefunction. This approach is used to study complex quantum systems with time-dependent interactions among their many parts and can manifest unexpected behavior.

It handles multi-dimensional integrals that arise in these problems.

- **Quantum Walks** they are used as an alternative to phase estimation. They use quantum superposition across states to model particle movement in applications such as light particle (boson) sampling. This approach can be used in many applications involving a stochastic process, such as searching, networks, and statistics.

It enables an exponential speedup.

5.1 GROVER UNSTRUCTURED SEARCH

This algorithm, focused on searching for an unsorted list of elements, demonstrated a quantum advantage in the early 1990s within the field of quantum mechanics.

The algorithm is defined to efficiently search through an unsorted database, which is an unordered list of N elements. It refers to the process of looking for a specific item within a collection of items that have no particular order or sequence, similar to finding a needle in a haystack. A visual analogy is shown in Figure 5.1.

The problem can be visualized as searching for an item in a box filled with unorganized assorted objects.

If you are looking for a specific object, you need to pull out objects one by one and look at them until you find the one you are looking for, or realize that it is not in the box at all.

The larger the list of objects, the longer it takes to find an element within it.

FIGURE 5.1   Grover algorithm visualization.

In a classical approach, the algorithm examines each item one by one, in a brute-force approach, systematically checking all possible candidates for the solution. It looks into each item iteratively, to check if it matches to the one it is looking for. In a list of N items, it will look into N/2 in the best scenario in average, and into all N items in the worst scenario, which is a *linear search that corresponds to a polynomial time complexity N.*

On the other hand, the quantum algorithm uses quantum superposition of all possible search states, representing all N possible items in the list simultaneously. It then applies amplitude amplification, which selectively enhances the probability of the correct item, making it more likely to be identified in the result measurement. It does this by iteratively applying a unitary operation, the Grover operator.

Algorithm improvement

This approach provides a quadratic polynomial speedup improvement, *significantly reducing the time required to complete a task. So, for example, if something takes 100 steps to complete, with the quadratic speedup it would take 10.*

5.2 SHOR NUMBER FACTORIZATION

This algorithm, focused on factorization to find prime numbers, demonstrated a quantum advantage in the early 1990s by utilizing the power of quantum mechanics.

The algorithm is defined to find all prime factors that compose a number and is solved through an iterative process that breaks the number down in steps. This process involves breaking down a composite number into its prime factors. A visual analogy is shown in Figure 5.2.

The problem can be visualized as a matryoshka doll, representing the number to be factored (composite).

The doll contains many nested smaller dolls, which are the prime factors. When simply examining the large doll, we do not know how many dolls it contains. The larger the number is (bigger the matryoshka doll), the more potential candidates (dolls) for prime factors are inside.

FIGURE 5.2 Shor algorithm visualization.

In a classical approach, the algorithm considers systematically all the potential prime factors in a brute-force approach, starting with the smallest prime number 2. It divides the composite number by the prime numbers. If the result is even, then the number is a factor, otherwise it is not. Each factor is subsequently divided by that prime number or the next one, until the result is 1, the smallest possible factor number. This process has an *exponential or sub-exponential time complexity* 2^N, which can become unfeasible to factor large numbers since it grows quickly.

On the other hand, the quantum algorithm performs a QFT and exploits quantum parallelism to find the modulus of a coprime reducing the complexity to $N^2 \log N^3$, *which is now polynomial.*

Algorithm improvement

This approach provides an exponential speedup *improvement drastically reducing* the time to complete the task. *So, for example, if we factor a 30-bit number, it takes* 2^{30} *which is* ~ *1 billion operations, instead the exponential speedup takes* ~ *10,000 operations.*

5.3 DEUTSCH–JOZSA ORACLE

This algorithm illustrated quantum advantage in the 1990s by efficiently determining the nature of a function represented by a 'black box' or 'oracle.' The algorithm provides a yes/no response faster than classical algorithms. It is a foundational algorithm that has inspired many algorithms, including search and machine learning.

The algorithm makes use of quantum superposition to determine whether a function is constant (producing the same outputs for all inputs), or balanced (producing 50% one and 50% zeros), using a minimum number of queries to the oracle. The algorithm initializes all qubits, except one, to 0. After querying the oracle and performing subsequent superpositions, the results reveal whether the nature of the function defined by the oracle is constant or balanced.

Algorithm improvement

Classical algorithms require 2 to $2^{n-1}+1$ queries to the oracle in the best case to check if the outputs are the same, while the quantum algorithm completes this task in 1 step, with significantly fewer queries.

5.4 BERNSTEIN–VAZIRANI DECODE

This algorithm is like the Deutsch-Jozsa Oracle, in the sense that it illustrated quantum advantage in the 1990s by identifying a hidden string encoded within a black-box function with minimum querying. It is a foundational algorithm that has also inspired algorithms like Grover's search.

The algorithm makes use of quantum parallelism and quantum interference to find in a hidden *string S* encoded in a black-box function $f(x)$. The algorithm initializes all qubits, except one, to 0, it applies a superposition and feeds them into $f(x)$. The black-box function flips all phase states where $f(x) = 1$ and applies another superposition that creates a constructive interference revealing the corresponding state of the hidden *string S* when it is measured and creating destructive interference for all other ones.

Algorithm improvement

Classical algorithms require N queries to find S, while the quantum algorithm requires only 1 query.

5.5 HARROW–HASSIDIM–LLOYD LINEAR SOLVERS

The Harrow–Hassidim–Lloyd (HHL) algorithms showcased a quantum advantage in the early 2000s by solving linear systems of equations exponentially faster than classical algorithms. They are foundational algorithms that have inspired many other algorithms, including machine learning and optimization.

The algorithms make use of the QFT and phase estimation to streamline matrix calculations. Quantum states are used to represent matrices and to estimate the specific properties of the matrices. Given a linear system of equations with two vectors, one known and the other unknown the goal is to find the value of the unknown vector.

Algorithm improvement

Classical algorithms require N^3 to find the solution, while the quantum algorithm takes $\log (N)^2$, where N is the number of variables.

5.6 QUANTUM METROPOLIS EQUILIBRIUM

This algorithm has demonstrated the potential for a quadratic speedup, and it is used to simulate complex physical systems and solve problems by employing principles of quantum mechanics. It is designed to sample from probability distributions that arise from a quantum Hamiltonian in systems with many possible configurations.

The algorithm leverages superposition to explore a high-dimensional state space and to efficiently sample from complex probability distributions. It employs phase estimation to calculate acceptance probabilities for all possible states within the desired distribution.

Algorithm improvement

Classical algorithms require $1/\delta$, where δ is the gap of the Markov chains, while the quantum algorithm takes $1/\sqrt{\delta}$.

CHAPTER 6

Applied Algorithms

S OME QUANTUM COMPUTING ALGORITHMS are used to either solve a practical problem or to perform a specific task or subroutine, bridging the gap between theoretical algorithms and their applicability in everyday life. Applied quantum computing algorithms go beyond theoretical purposes to address real-world problems or applications, offering tangible advantage over classical computing.

These applied quantum algorithms and techniques represent today's latest quantum computing research and development to solve problems across various scenarios. However, the practical implementation of these algorithms require advancements in error correction techniques, some of them requiring fault tolerant quantum computers.

Some of the algorithms described are considered Variational Quantum Algorithms (VQAs), are a class of quantum algorithms that use a hybrid quantum-classical approach, that due to their adaptive nature, they are well suited to handle the constraints of near-term quantum computers that have noise and can show a quantum advantage.

Examples of Variational Quantum Algorithms that are described on this section are those based on iterative calculations (optimization) and artificial intelligence.

The "variational" aspect refers to the fact that these algorithms rely on the optimization of parameters within a quantum circuit, guided by classical computation. It involves a quantum circuit with tunable parameters that are adjusted by the classical computer to minimize a cost function

DOI: 10.1201/9781003302674-8

that is calculated using the quantum computer. This process iteratively improves the solution to the encoded problem.

The process is shown in Figure 6.1.

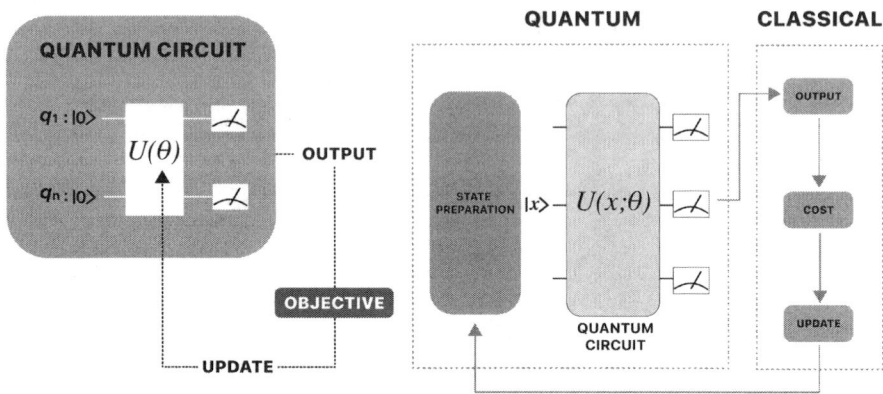

FIGURE 6.1 VQA Illustrative conceptual diagram.

6.1 VARIATIONAL QUANTUM EIGENSOLVER (VQE)

The VQE algorithm is used to analyze the electronic Schrödinger equation, a fundamental tool for calculating and understanding the electronic structure of molecules and atoms, with an expected **exponential speedup**.

Solving the electronic Schrödinger equation is challenging due to the intricate interactions among electrons within the system, constituting a "many-body" problem. Solutions to problems in quantum chemistry and condensed matter physics are often found through approximation.

Algorithm improvement

VQE algorithm attempts to find a wavefunction that minimizes the energy expectation of a system using the variational principle from quantum mechanics, which states that the energy of a trial wavefunction is always greater than or equal to the true ground-state energy. This approach enables VQE to efficiently estimate the ground-state energy of a complex quantum system.

6.2 QUANTUM AMPLITUDE ESTIMATION (QAE)

The QAE algorithm's goal is to determine the probability of measuring a specific outcome in a stochastic system where there is an element of uncertainty, that can be analyzed by performing random sampling to simulate the system behavior numerically with almost **quadratic speedup**.

Solving Monte Carlo methods is complicated, because properly simulating the behavior of the system produces many random samples to estimate the expected output value of a randomized algorithm. Due to computational needs, the results are averaged to find an approximation.

Algorithm improvement

QAE algorithm is used to estimate probabilities using superposition, amplitude amplification, phase estimation, and quantum oracle techniques to estimate the amplitude, which corresponds to the probability of measuring the specific outcome encoded in the quantum states.

6.3 QUANTUM APPROXIMATE OPTIMIZATION ALGORITHM (QAOA)

The QAOA is designed to find an optimal or near-optimal solution from a finite set of possibilities, where each solution has an associated cost that requires optimization over a set of binary variables representing the combinatorial problem, with an expected more **effective approximate solution**.

Solving optimization problems involves exploring many variables that can be heterogeneous, have nonlinear relationships, and grow exponentially with the problem size. They have multiple minima (value smaller than the neighboring points) and several simultaneous constraints, which makes them challenging to solve.

Algorithm improvement

QAOA makes use of quantum superposition, simultaneously evaluating multiple potential solutions that have been encoded into strings of binary variables. It encodes and evaluates the cost function using an iterative approach to refine and identify the best solution.

6.4 QUADRATIC UNCONSTRAINED BINARY OPTIMIZATION (QUBO)

As a form of mathematical minimization problem with a specific criterion for making yes/no decisions with no explicit constraints, QUBO admits at least one optimal solution but does not necessarily need to have multiple optimal solutions, allowing **multiple results**.

Expressing the optimization problem in QUBO form has significant drawbacks due to constraint modeling using penal quadratic variables, which greatly expands the solution space, increasing the complexity of the problem and limiting the types of problems that can be solved well.

Algorithm improvement

QUBO algorithm maps the variables into a quantum format and a Hamiltonian form, which makes it easy to analyze the expected values of the parameters, enabling us to find solutions that satisfy the constraints without forcing them to hold.

6.5 QUANTUM DIFFERENTIAL EQUATION (QDE)

The QDE algorithm is used to solve differential equations, either ordinary or partial differential equations; these are fundamental tools for solving functions of several variables and can model continuous processes with an **expected exponential speedup**.

Solving differential equations, such as Navier–Stokes equations, can be very complex because they involve multiple independent variables. They model complex systems, which makes it mathematically difficult to solve them and challenging to find accurate solutions.

Algorithm improvement

A QDE encodes information into a quantum high-dimensional state, encoding exponentially long vectors for the development of quantum linear systems, which can then be measured to efficiently extract the solution using the QFT.

6.6 QUANTUM ARTIFICIAL INTELLIGENCE (QAI)

Quantum computing is expected to accelerate Artificial intelligence, by enabling to extract complex patterns from data, a crucial aspect of AI technologies. Solving machine learning models is complicated because they may need analyzing large and high-dimensional datasets, which require a long time to process and many data points for training to achieve high precision; this may not be achieved in a practical timeframe.

An example is Large Language Models, which are at the forefront of current AI solutions, involve processing an enormous amount of text data and configurations to construct sentences. This task is computationally intensive, and the data required to train such model is huge. Some of the machine learning techniques used to process and analyze data to find patterns are neural networks, support vector machines, principal component analysis, Boltzmann machines, and classifiers. Quantum versions are expected to bring **quadratic and exponential speedups.**

Algorithm improvement

The algorithms are used to train and create models to generate correlations between variables and make the output closer to the quantum features. They transform matrices into an exponential 2^n-dimensional complex vector space, mapping data to a large feature space for improved exploration. It has the ability to handle high-dimensional data with fewer parameters and perform computations that are classically intractable.

- **Quantum Neural Networks (QNNs):** QNNs are made of layers of quantum gates that act as neurons and process information through complex transformations. These networks exploit quantum parallelism for information processing. The gates parameters correspond to the neural network weights are optimized to perform the tasks.

- **Quantum Support Vector Machine (QSVM):** QSVM maps classical data into a high-dimensional quantum feature space and then separates it through a hyperplane. It uses a quantum kernel estimation technique that is key to compute the inner product in the feature space implicitly, without needing to compute the feature map explicitly.

- **Quantum Principal Component Analysis (QPCA):** QPA encodes data into quantum states leveraging quantum phase estimation and quantum singular value decomposition to identify the most significant features of a dataset. This technique is used to reduce dimensionality more efficiently in machine learning which is useful to handle large datasets.

- **Quantum Boltzmann Machine (QBM):** QBMs use quantum systems to represent the Boltzmann distribution that is a type of stochastic recurrent neural network. It allows to represent the complex probability distributions with fewer resources and through superposition and entanglement explore the solution space when learning complex patterns and correlations in data.

- **Variational Quantum Classifier (VQC):** VQC are used for classifying data, by encoding it into a quantum state mapping it, using an elementary circuit architecture to build the full variational circuit. Once it creates the feature mapping, it adjusts the parameters of the quantum gates, that are constantly optimized during the training process, to minimize a cost function. At the end it measures the output to determine the class.

6.7 ALGORITHMS INTERRELATION

In quantum computing, certain algorithms and fundamental techniques serve as crucial components or 'building blocks' for more complex algorithms. These foundational elements are utilized to enhance the capabilities of other, more advanced algorithms. This interconnected nature of algorithms is crucial in understanding how simpler processes contribute to the functionality of more complex quantum problems.

Figure 6.2 visually demonstrates the integration of these foundational components into different algorithms, highlighting their interconnectedness and importance in advancing complex quantum computing operations.

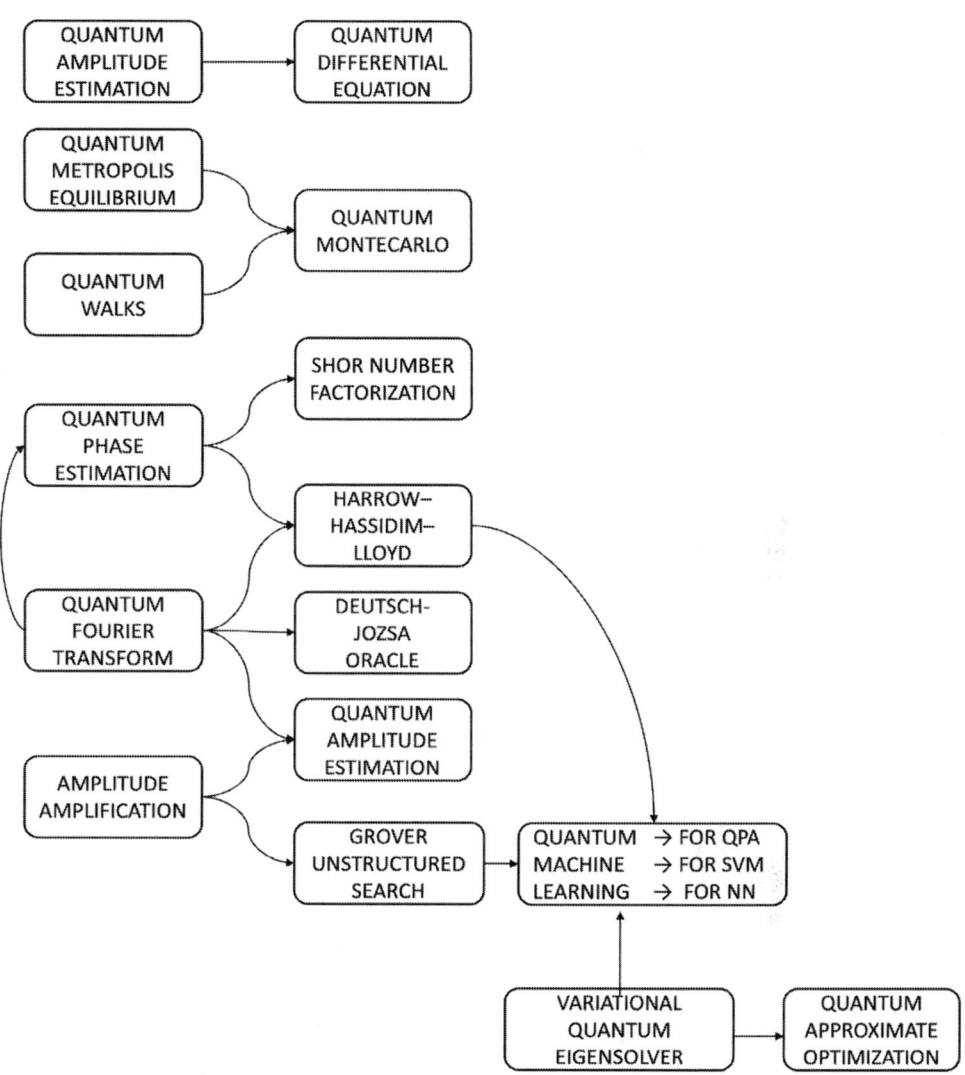

FIGURE 6.2   Algorithm interrelation.

CHAPTER 7

Problem Categorization

Q UANTUM COMPUTING IS A promising technology that can offer efficiency in approaching mathematical problems and provide business value by saving money in current high-performance computing usage, solving a problem more efficiently, or addressing a problem that cannot currently be solved.

Quantum algorithms introduce computational value to the underlying formulas used in different industry problems, and we can group them into four problem categories, each of them applied to a specific industrial field.

The different categories are shown in Figure 7.1.

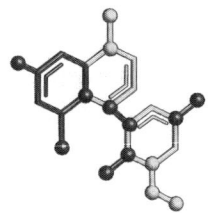

Chemical processes

Stochastic simulations

Objective optimization

Artificial intelligence

FIGURE 7.1   Problem categories.

DOI: 10.1201/9781003302674-9

Table 7.1 describes each of the categories, some examples, and formulas.

TABLE 7.1	Problem Categories Description
Chemical processes	**Category definition:** to analyze how chemical structures make up materials, molecular interactions and properties in ways that are essential to our very existence and explain our surroundings.
	Applicability examples: drug design and chemical reactions, which are used in multiple industries such as pharma, oil and gas, and materials.
	Some formulas: are molecular dynamics, density functional theory, coupled cluster methods, reaction route search, and vibration analysis.
Stochastic simulations	**Category definition:** a process forecasting the evolution in time of a random phenomenon. It creates simulation models that are data-dependent based on real-world information that is sampled randomly.
	Applicability examples: risk assessment, analyzing impact of variability under different conditions, simulating, modeling, and predicting systems.
	Some formulas: are Monte Carlo simulations, Navier–Stokes, Black–Scholes, Bayesian modeling, or Differential Equations, or Markov Chains.
Objective optimization	**Category definition:** to find the best solution, maximum or minimum, from all feasible solutions. This approach has a set of decision variables, an objective function, bounds on the decision variables, and constraints.
	Applicability examples: routing, scheduling, finding the best path in a network, combining elements, or designing optimal structures.
	Some formulas: are linear programming, network planning, graph queuing, minimum spanning tree algorithm, or gradient descent.

(Continued)

TABLE 7.1 (*Continued*) Problem Categories Description	
Artificial intelligence	**Category definition:** the capability of a computer system to mimic human cognitive functions. Machine learning applications use mathematical models to help computers learn without direct instruction.
	Applicability examples: image and speech recognition, natural language processing, recommendations, predictive analytics, or fraud detection.
	Some formulas: are supervised, unsupervised and semi-supervised learning, reinforcement learning, classification, and neural networks.

7.1 CHEMICAL PROCESSES

This category is focused on analyzing the atomic and electronic properties to simulate the behavior of quantum systems; e.g. to understand the color of a substance. It seeks to understand chemical interactions in processes such as catalysis or molecular design that are present in the design of engines, fertilizers, fuel cells, drugs discovery, reservoirs, carbon capture, materials, or energy sources.

7.1.1 Classical Approach

In a classical approach, it is important to calculate all combinations of electrons on molecular orbitals. This problem is difficult to address because of the massive number of electron arrangements (10 to the power of 300) for any given atomic arrangement, and it is impossible to address in a step-by-step fashion. There are approximation methods, that can reduce the computational cost and complexity, making it feasible to perform computations on large and complex molecules. However, the shortcoming of these methods is decreased accuracy.

7.1.2 Business Impact

According to data from recent quantum computing market analysis by McKinsey & Company, BCG, and interviews with SMEs, use cases involving stochastic simulations will mostly have a significant business impact in the short term. This impact turns into a mostly disruptive one in the long term after 2030, once the technology is mature enough to show its viability.

Table 7.2 shows the potential long-term business impact for some use cases that fall under this category.

TABLE 7.2 Chemical Processes Long Term Business Impact (Non-exhaustive)		
Disruptive ~ US\$B [60–90]	**Significant ~ US\$B [30–60]**	**Incremental ~US\$B [0–30]**
• Molecular properties • Aerospace • Supply chain • Chemical catalysis • Drug candidates	• Efficient fertilizer • EV batteries • Chemical catalysis • Improve materials • Clean energy • Radio material design	• Reservoir simulations • Carbon capture • Engine design

7.1.3 Quantum Approach

In a quantum algorithm approach, we try to get a more accurate microscopic understanding. We encode the electron states as qubits and apply quantum gates to simulate the interactions between electrons to extract accurate molecular energy levels. In those cases that the electron-electron interactions are dominant instead of weak, a quantum computer can accelerate the process exponentially.

Quantum algorithms usually used are Variational Quantum Eigensolver, Quantum Phase Estimation, or Quantum Walks, amongst others.

7.1.4 Specific Example

The challenge of protein folding involves calculating the lowest energy state of a protein's structure, given its amino acid sequence. This is a highly complex problem due to the vast number of possible configurations a protein can take.

Classical computers can simulate these processes, but the computational cost grows exponentially with the protein complexity, making it infeasible for large proteins.

Quantum computers, however, can potentially explore multiple configurations simultaneously due to quantum superposition and apply quantum entanglement to find correlations between different parts of the protein. This could allow them to evaluate a vast number of possible protein conformations more efficiently than classical computers. Quantum algorithms represent the problem into energy states, simplifying the problem and reducing the number of steps, as it can be seen in Figure 7.2.

Classical approach

Quantum approach

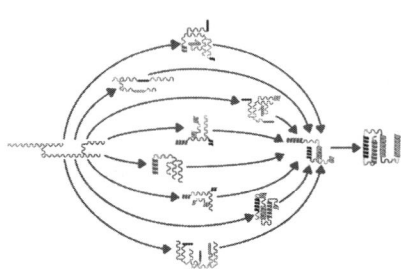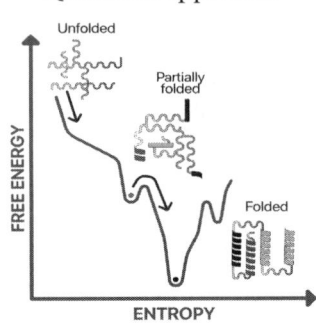

The classical approach explores a set of possible protein shapes to predict how they will move and fold over time.

A quantum approach calculates the energy states and entropy changes in folding to speed up the calculation.

FIGURE 7.2   Problem example: protein folding.

7.1.5 Simplified Analogy

To understand the difference in how quantum computing tackles problems, we explain an over-simplified analogy to illustrate how classical and quantum approach algorithms solve problems, using a color shade example shown in Figure 7.3.

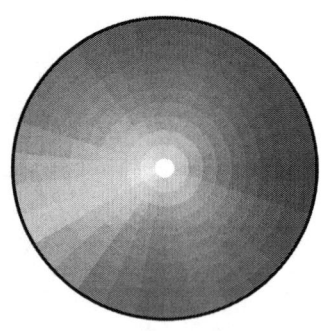

Problem definition

How to find a perfect shade of paint to match a specific color. There are several colors to choose from and mix to find the desired shade. The more complex the shade is, the bigger amount of color combinations there is to mix. In the palette there are base colors and a basic idea of how to mix them to approach the finding of desired shade.

FIGURE 7.3   Chemical processes simplified analogy.

(*Continued*)

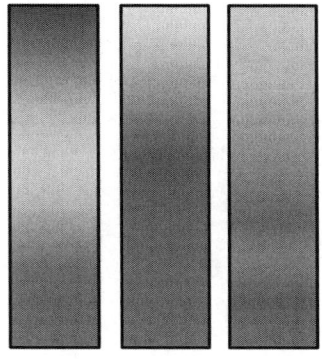

Classical approach

Uses a different set of palettes to iteratively mix different colors to find the desired shade. If the shade is not found with the chosen palette, then another palette must be chosen, and the mixing process starts again. This process is repeated iteratively until the right shade is found which is time-consuming.

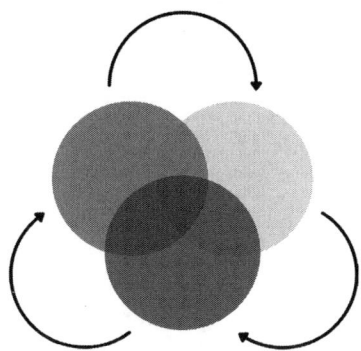

Quantum approach

Uses smart palette that can rotate the different base colors to use for mixing. This rotation facilitates the process by allowing simultaneously change of the proportion of the base colors to get closer to the perfect match when creating different color combinations.

This rotating process is like the VQE algorithm's method of optimizing parameters to minimize the energy of a quantum system.

FIGURE 7.3 (*Continued*)   Chemical processes simplified analogy.

7.2 OBJECTIVE OPTIMIZATION

This category is focused on the selection of a best element that can be either minimizing or maximizing a value from a set of available alternatives based on some constraints. It seeks to analyze the best feasible solution of a mathematical complex model making better use of available data that is applied in combinatorial problems to solve: electric power distribution, financial portfolio management, spectrum management, scheduling, transportation, or space layouts.

7.2.1 Classical Approach

In a classical approach, we investigate all the potential solutions using a "brute-force" approach, which means trying each one until a solution is found. This can become too difficult to solve if the problem size is large with multiple constraints and variables to calculate. If the number of variables is large and creates too many combinations to analyze, in the order of higher than 1,000 combinations, it cannot be solved in a reasonable amount of time. There are many ways to overcome these restrictions such as heuristic approaches, solvers that are tailored to an existing problem, and simplified models, indirect representation, or aggregations that seek to find approximate solutions that are not fully accurate.

7.2.2 Business Impact

According to data from recent quantum computing market analysis by McKinsey & Company, BCG, and interviews with SMEs, use cases that involve stochastic simulations will mostly have an incremental business impact in the short term that will turn into a mostly significant one in the long term after 2030, once the technology is mature enough to show its viability.

Table 7.3 shows the business impact breakdown in the long term for some examples use cases that fall under this category.

TABLE 7.3 Objective Optimization Long Term Business Impact (Non-exhaustive)		
Disruptive **~ US$B [60–90]**	**Significant** **~ US$B [30–60]**	**Incremental** **~US$B [0–30]**
• Routing flow • Transport efficiency	• Air cargo load • Seed biology • Energy commitment	• Risk management

7.2.3 Quantum Approach

In a quantum algorithm approach, we typically represent the optimization problem as the equivalent problem of finding the lowest energy state (known as the ground state) of a physical system. Using quantum interference, superposition, and entanglement, the physical system can explore exponentially more solutions configurations instead of analyzing one solution at a time.

Quantum algorithms usually used are Quantum Approximate Optimization Algorithm, Quadratic Unconstrained Binary Optimization, or HHL, among others.

7.2.4 Specific Example

The challenge of route optimization is that the algorithm searches through all possible combinations of routes to find the most efficient one according to specified criteria, such as distance, time, or cost, one path at a time, solving it sequentially which can take very long as the number of routes grows. Quantum computers, however, can potentially explore multiple route paths simultaneously due to quantum superposition and entanglement to encode the constraints configuring all potential routes at once. This could allow them to evaluate a vast number of possible route conformations more efficiently than classical computers. Quantum algorithms map the problem into energy states, simplifying the problem and reducing the number of steps, as shown in Figure 7.4.

<table>
<tr><td>Classical approach</td><td>Quantum approach</td></tr>
</table>

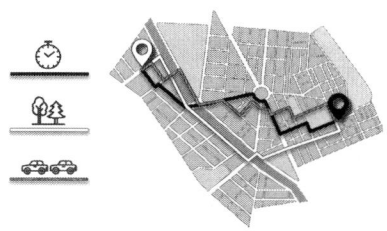
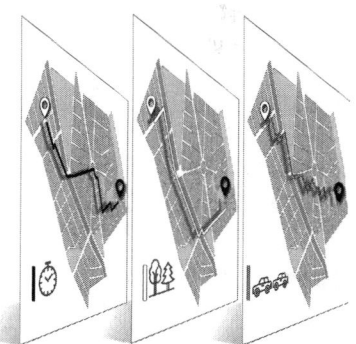

In a classical approach we sequentially examine one by one all possible paths to find the one we are looking for.	In a quantum approach we can view all the routing options in parallel so we can choose the one we are looking for faster.

FIGURE 7.4   Problem example: route optimization.

7.2.5 Simplified Analogy

To understand the difference in how quantum computing tackles problems, we explain an over-simplified analogy that helps to visualize the difference on how classical and quantum approach algorithms solve problems using exiting a maze example shown in Figure 7.5.

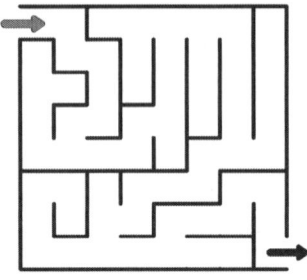

Problem definition

Finding the exit in a maze that has multiple paths that lead nowhere. In the maze there are some paths that lead to the exit, but we do not know which is the right one. The larger the number of paths the maze has, the more complicated finding the exit gets. In the maze, there is as intuition of the direction towards the exit.

Classical approach

As we go through the maze, we need to try one path at a time. So, each time there is a bifurcation, we choose one direction. If we hit a dead end, we must go back to the bifurcation point and try the other direction. This process must be repeated iteratively until we find the maze's exit.

Quantum approach

As we go through the maze, we can simultaneously explore all paths at once. We have the capacity, once we reach a bifurcation, to make a "copy" of ourselves to explore both directions at the same time and provide feedback on the result. This is repeated iteratively until we exit the maze.

FIGURE 7.5 Objective optimization simplified analogy.

This multiple path exploration process is similar to the QAOA algorithm method of multiple solution analysis and probability adjustment.

7.3 STOCHASTIC SIMULATIONS

This category is focused on modeling uncertainty and variability in a chaotic environment that requires mathematical statistical analysis to predict potential outcomes. Even if we cannot predict an individual outcome in a random process, we can still gain a valuable statistical understanding. We can study the variance of time series based on historical data to estimate the probabilities that are present in many applications usually solved with complex differential equations to models random process.

7.3.1 Classical Approach

In a classical approach, when simulating the stochastic process that represents a system, it generates many random variables to generate the probability distribution of all possible outcomes that mimic real-life results. It calculates over the results and averages them estimate the desired quantity. The accuracy of the estimation improves with the number of samples, for example increasing them by a factor of a 100 to decrease the error by a factor of 10. Running a significant number of simulations may require substantial computational power and time, especially when dealing with complex systems and many iterations.

7.3.2 Business Impact

According to data from recent quantum computing market analysis by McKinsey & Company, BCG, and interviews with SMEs, cases that involve stochastic simulations will mostly have an incremental business impact in the short term that will turn into a mostly significant one in the long term after 2030, once the technology is mature enough to show its viability.

Table 7.4 shows the potential long-term business impact for some use cases that fall under this category.

TABLE 7.4   Stochastic Simulation Long Term Business Impact (Non-exhaustive)

Disruptive ~ US$B [60–90]	Significant ~ US$B [30–60]	Incremental ~US$B [0–30]
• Underlying AI algorithms	• Money laundering	• Object detection in autonomous driving and medical imaging

7.3.3 Quantum Approach

In a quantum algorithm approach, similar to Monte Carlo methods, differential equation-style problems are solved more efficiently. Instead of using numerous repeated random samplings analyzing all the simulations and averaging the solutions, we leverage selective amplification to converge to an answer faster, generally with a polynomial speedup.

Quantum algorithms usually used are Quantum Amplitude Estimation, Grover Unstructured Search, or Quantum Fourier Transform, among others.

7.3.4 Specific Example

Monte Carlo simulations rely on repeated sampling to obtain numerical results by generating random numbers. It follows the specified probability distributions for the uncertain variables, and represents the possible set of conditions out of the many that could occur. Determining when a simulation has run enough iterations to be reliable, can be a complex task and converging to an accurate solution is challenging. Quantum Monte Carlo simulations have the potential to converge to the solution faster. The difference between classical and quantum is seen on Figure 7.6.

Classical approach

Quantum approach

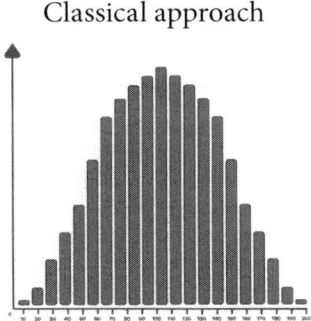 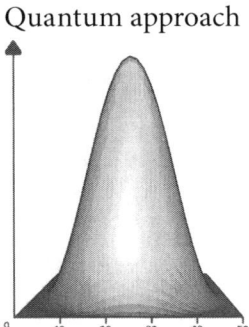

In a classical approach, we run many simulations and aggregate the outcomes estimating the distribution mean.

In a quantum approach, we improve the approach with a 3-dimension distribution increasing the number of simulations.

FIGURE 7.6   Problem example: probability distribution.

7.3.5 Simplified Analogy

To understand the difference in how quantum computing tackles problems, we explain an over-simplified analogy that helps to visualize the difference on how classical and quantum approach algorithms solve problems using a polling example shown in Figure 7.7.

Problem definition

The techniques used to predict the opinion, preferences, or attitudes about a specific topic involve surveying a sample of people that are representative of the wider population.

The wider the population is in terms of diversity and variability, the more important it is to have a large amount of prior data, which in turn aggregates results and analyzes them.

Classical approach

To aggregate the polling data, after collecting the responses from the sampled population, they must be averaged to for example calculate the percentage of respondents who favor a certain topic that can provide an estimate of the overall opinion. Estimation accuracy increases linearly with the number of samples but so does the resources and runtime need.

FIGURE 7.7 Simplified analogy stochastic simulation.

Quantum approach

To calculate the average, it encodes opinion samples into qubits, which store and correlate large amounts of data, enabling a more efficient analysis. It computes based on state probabilities, offering a detailed perspective on people's intentions, leading to faster, more precise, and better computational outcomes.

Encoding data into quantum states is like QAE algorithm use of probability amplitudes to estimate properties derived from them.

7.4 ARTIFICIAL INTELLIGENCE

This category is focused on evaluating data to make predictions, find patterns or correlations, build assumptions, classify information, without explicit instructions, based on its learning and thinking. It seeks to make decisions based on vast amount of historical data used to train it, that can be used to automate tasks, analyze data, interpret human language, recognize images, and make recommendations.

7.4.1 Classical Approach

In a classical approach, the process includes definition of the process to be solved, collect datasets with historical or newly generated data that must be processed to train the algorithm. Based on the problem to be solved and the type of data, a different model will be chosen to train and teach the algorithm, which will then be tested and fine-tuned. Large heterogeneous dataset in nature, data that is structured, unstructured, multimedia, and text, need different processing techniques and can complicate the training model making it hard to find relationships between features and outcomes.

7.4.2 Business Impact

According to data from recent quantum computing market analysis by McKinsey & Company, BCG, and interviews with SMEs, use cases that involve artificial intelligence, will mostly have an incremental business impact in the short term that will increase to some significant one in the long term after 2030 due to the skepticism on quantum computing viability for this category.

Table 7.5 shows the business impact breakdown in the long term for some examples use cases that fall under this category.

TABLE 7.5 Artificial Intelligence Long Term Business Impact (Non-exhaustive)		
Disruptive **~ US$B [60–90]**	**Significant** **~ US$B [30–60]**	**Incremental** **~US$B [0–30]**
• Crystal structures • Protein pathology • Just-in-time logistics • Weather forecast	• Satellite imaging • Product pricing • Grid operation	• Predictive maintenance • Service quality • Irregular operations • Object detection

7.4.3 Quantum Approach

In a quantum algorithm approach, we enhance machine learning algorithms by analyzing classical data on a quantum computer. The quantum algorithm uses amplitude encoding to store 2^N solutions with N qubits and reduce the needed resources exponentially. As the problem size increases, the required resources grow at a logarithmic rate, rather than increasing in proportion to the size of the input.

Quantum algorithms usually used are Quantum Machine Learning, Quantum Neural Networks, or Quantum Support Vector Machine, among others.

7.4.4 Specific Example

Artificial intelligence relies on finding patterns in data by analyzing correlations between data points looking into its different features, which are the different variables that it has, that are used to train the model. Features must be prepared extracting them from the data source, they must be transformed to a format which can be analyzed. Finally, the most relevant features must be identified, and selected to be used in a model. It is complex to balance the number of features, called dimensions, non-linear interdependencies, with the model complexity to find an adequate dividing classification plane.

The difference between classical and quantum is shown in Figure 7.8.

Classical approach   Quantum approach

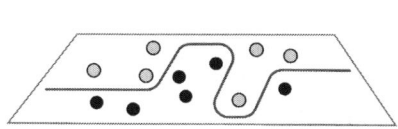   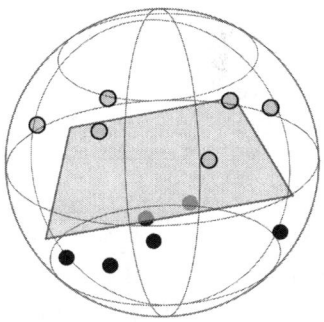

In a classical approach the algorithm finds the optimal linear separation of different classes by identifying the data points closest to it.

In a quantum approach the algorithm uses a 3-dimension feature space that splits data regardless of its complexity, creating a more precise AI model.

FIGURE 7.8   Problem example: classification process.

7.4.5 Simplified Analogy

To understand the difference in how quantum computing tackles problems, we explain an over-simplified analogy that helps to visualize the difference on how classical and quantum approach algorithms solve problems using the idea of finding a book shown in figure 7.9.

Problem statement

Books are usually categorized in libraries by analyzing the genre they belong to and placing them on different shelves that are labeled with the genre type. To facilitate the search, it is important to classify the book as accurately as possible. This classification can be hard if the book has many overlapping characteristics that makes it hard to distinguish the genre.

Classical approach

To look for a book, the librarian must look into the books one by one, checking certain features to decide which genre the book belongs to. After reviewing enough books, the librarian obtains a good intuition to decide how to make this distinction. Once a book genre is identified, it is placed on a specific shelve close to similar genre books.

FIGURE 7.9 Simplified analogy artificial Intelligence.

Quantum approach

When the librarian opens one book, other books will automatically open to pages with content that is directly related, making it easy to understand which books belong to the same category. As the quantum algorithm represents the book genre into quantum states and links all those related, books do not need to be placed physically together. This exponentially improves the search and answers to an AI query.

CHAPTER 8

Quantum Computing Risk

CRYPTOGRAPHY IS AN ANCIENT technique used for encrypting information to keep it secret and prevent it from being read by an unauthorized party. At its core, cryptography involves taking readable data (plaintext) and using a mathematical process (algorithm) along with a secret key to transform it into an encoded version known as ciphertext. To decipher the ciphertext, a secret key must be used. In the case of a symmetric cipher the same key is used for encryption and decryption, as we explain in this section.

The practice of encrypting messages is over 4,000 years old, with the term "cryptography" deriving from the Greek words "Krypto" (hidden) and "Graphos" (writing), meaning "hidden writing". The field advanced in post-World War II using complex mathematical processes to create encryption techniques. This led to today's standard encryption algorithms such as the Advanced Encryption Standard (AES), Rivest–Shamir–Adleman (RSA), and Elliptic Curve Cryptography (ECC), each detailed in Table 8.1.

Secure cryptographic schemes are essential for protecting government and military communications, financial and banking transactions, confidentiality of medical data and healthcare records, storage of personal data in the cloud, or restricted access to confidential networks. Quantum computers pose a significant threat to current cryptographic methods, as their advanced capabilities could render today's math-based encryption obsolete. This jeopardizes the security of data, whether stored or transmitted,

DOI: 10.1201/9781003302674-10

including everyday encrypted information like emails, phone calls, passwords, cloud storage, financial transactions, digital certificates, and medical records.

TABLE 8.1	Standard Encryption Algorithms		
Algorithm	**Keys**	**Applicability**	**Requirements**
AES	Symmetric key: same key is used to encrypt and decrypt.	Encryption of data at rest, stored on a hard drive, or in a database.	Straight forward but its security depends on the key's length.
RSA	Asymmetric key: uses a public key to encrypt and a private key to decrypt.	Encryption of data for internet transmission.	Long keys make it more secure but also computationally heavier.
ECC	Asymmetric key: uses a public key to encrypt and a private key to decrypt.	Encryption of data for digital signatures.	High security with shorter keys, making it computationally lighter.

Although quantum computing may not be fully working and does not currently pose a risk to cyber-security schemes, hackers may be capturing and storing today's data to decrypt it tomorrow. This is called the 'harvest now, decrypt later' approach, which could allow an eavesdropper to successfully crack an encryption scheme retroactively.

There are many opinions as to when quantum computers will be able to break current-day cryptography standards. Michele Mosca, deputy director of the Institute for Quantum Computing at the University of Waterloo, estimates that by 2030, more than half of today's cryptographic schemes will be obsolete. This security threat leads to what is called the "year to quantum", or the so-called Y2Q effect in which new cryptography schemes are needed to stop a potential attack by an adversary with a quantum computer. This is also referred to as Q-Day, the time when quantum computers will be able to decrypt the codes safeguarding our digital information making it vulnerable.

There is a global effort to mitigate this security threat in two fronts:

1. Implement secure cryptographic schemes to protect data from both classical and quantum threats, ensuring it is kept safe and transmitted securely.

2. Create a post-quantum transition plan to update to the new cryptography schemes and new standards seemingly without creating disruptions.

Governments and international standards organizations are encouraging the creation and application of post-quantum cryptography as soon as possible.

- NIST is working on quantum-resistant cryptographic algorithms.

- ETSI has a Working Group for quantum-safe implementation considerations and protocols.

- NSA has released the Quantum Computing Commercial National Security Algorithm Suite.

8.1 QUANTUM CRYPTOGRAPHIC SCHEMES

There are different security approaches making it quantum resistant, the main difference is how it is built, one approach is based on quantum mechanics to create cryptographic schemes and the other is based on mathematical methods.

There are three main different security approaches:

1. Post-quantum cryptography (PQC) is based on conventional mathematical cryptographic systems that are extremely complicated and secure against both quantum and classical computers. It interoperates with existing communications protocols and networks. It is also called quantum-safe and quantum-proof algorithms.

(Continued)

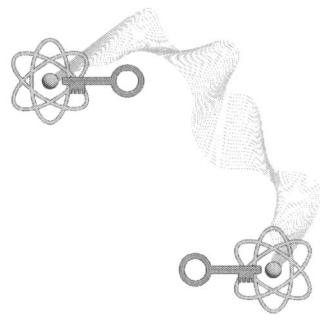

2. Quantum secure cryptography (QSC) relies on quantum mechanics properties instead, which makes eavesdropping impossible, as the system can detect if the key has been corrupted and discards it.

It exploits the no-cloning theorem and how a system cannot be observed without changing or disturbing it. An example is quantum key distribution (QKD) which secures communication with a new cryptographic protocol.

3. Quantum Random Number Generators (QRNGs) use quantum mechanics to generate genuinely random numbers, avoiding the issue of pattern repetition seen in conventional pseudo-random number generators. This technology is now available in commercial hardware chips, finding applications across multiple industries such as automotive, mobile devices, and IoT devices.

8.2 QUANTUM-SECURE CRYPTOGRAPHY QKD

Quantum Key Distribution (QKD) is the most common type of quantum cryptography, and it provides a way of distributing and sharing secret keys for cryptographic protocols. QKD originated in the 1970s with Stephen Wiesner's concept of quantum conjugate coding, which allowed for secure message transmission by making the reading of one message destroy the others. His foundational work, published in 1983, paved the way for further developments by Charles H. Bennett and Artur Ekert, who established key protocols for secure quantum communication, prepare and measure based (BB84) and entanglement based using the violation of Bell's theorem (E91), that makes it possible to receive and decode one variable but not both simultaneously.

QKD protocols utilize quantum systems, such as photons, to create a private key shared by two parties, allowing for the safe transfer of data either through direct optical fibers or via satellite links. The goal of QKD is to encrypt data for safe storage or transmission using two conjugated variables that cannot be switched, based on Heisenberg's uncertainty principle. Because of this, the cable doesn't need to be physically secured. Each photon will have its own randomized quantum state. Should someone be eavesdropping, both transmitter and receiver will be able to detect it because it's impossible to observe a quantum state without also affecting it, which is why QKD are considered un-hackable.

Figure 8.1 shows the QKD protocol scheme between A and B, with all the involved parts:

- laser beam to transmit bits through photons that carry key values.

- different filters, one to encode by A and other to decode by B.

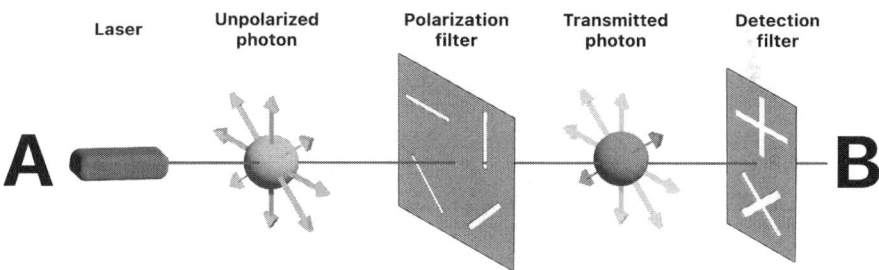

FIGURE 8.1   QKD transmission scheme.

Transmitter A uses a polarization filter to encode the value into four possible states: horizontal, vertical, diagonal left, or diagonal right. Receiver B uses a random detection filter with straight and diagonal lines to measure the bit's value, and then checks if basis matches A's. If there is a match, the value is kept. Eavesdropping (ear dropping) is impossible since it changes the qubit state, and therefore, the reading state would not coincide. This scheme is shown in Figure 8.2.

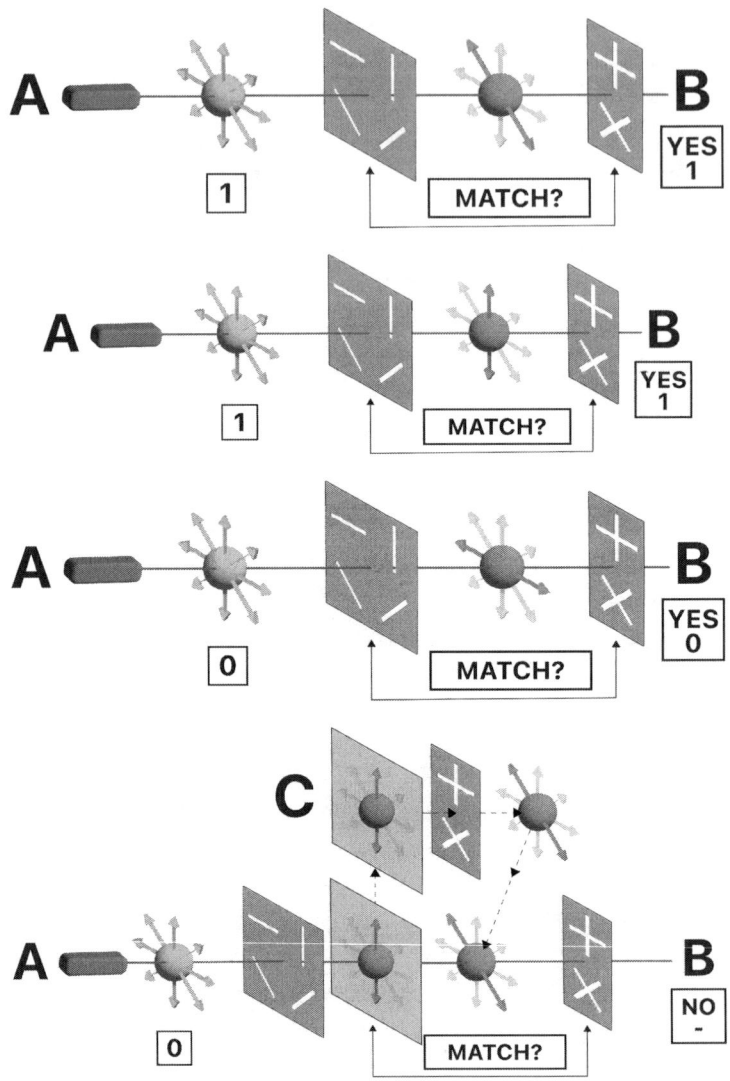

FIGURE 8.2   Example of complete QKD scheme.

8.3 POST-QUANTUM CRYPTOGRAPHY ALGORITHMS

Quantum computers can break many public-key cryptosystems, including the widely adopted RSA. Some estimates that state that using quantum algorithms an RSA asymmetric 2048-bit length key code can be broken in just 8 hours, versus the estimated 300 trillion years for a classical computer. For this reason, there is a need to update existing cryptographic standards with Post-Quantum Cryptography to ensure the security of cyber systems in the new age of quantum computing. Post-quantum cryptography research focuses on six approaches to create cryptographic functions that are believed to be hard for quantum computers to break.

Table 8.2 shows the approaches to create secure cryptography.

TABLE 8.2   Secure Cryptography Approaches
Lattice-based, use regular mathematical grids of points in multi-dimensional space.
Multivariate, based on solving multivariate polynomial equations over a finite field.
Hash-based, transform input data into a random fixed-size character string.
Code-based, involves problems from coding theory, such as error-correcting codes.
Super singular points between elliptic curves keeping its algebraic structure.
Symmetric keys depend on large, identical secret encryption and decryption keys.

In 2016, NIST initiated a contest to identify cryptographic methods resistant to quantum attacks, aiming for a new standard. Out of the initial 82 submissions, four were selected after several narrowing rounds. The winners were selected based on their robust schemes and practicality shown on Table 8.3.

TABLE 8.3 PQC Selected Algorithms	
1. CRYSTALS-Kyber key encapsulation mechanism.	**Lattice-based**
2. CRYSTALS-Dilithium is for digital signatures.	**(idem)**
3. FALCON is a lightweight digital signature.	**(idem)**
4. SPHINCS+ for document signatures	**Hash-based**

8.4 QUANTUM SAFETY STRATEGY PLAN

Organizations must update their cryptographic infrastructure to protect data and be prepared for the quantum threat. This involves ensure resilience against quantum threats and securing data long-term following the five steps shown on Figure 8.3.

FIGURE 8.3   Quantum safe readiness framework.

1. **Conduct a security audit** to identify data at risk, catalogue all data, assess the storage devices, evaluate data's critical nature, and determine whether it is static or dynamic.

2. **Monitor PQC standards** to constantly update to new quantum-resistant encryption algorithms, that ensure data is protected with the latest technology.

3. **Create crypto-agile systems** adding an encryption scheme that allows to seamlessly propagate any updates to all linked infrastructure maintaining operationality.

4. **Prepare all employees** by cultivating a culture of understanding post-quantum impact and up-skill all security personnel to strengthen the organization readiness.

5. **Implement transition plans** to migrate infrastructure to post-quantum cryptography either proactively or reactively retrofitting later, prioritizing what data to protect.

This strategic plan will enable companies to transition to post-quantum cryptography and effectively strengthen its defenses, detection, and response capabilities to reduce risk. It is important to ensure that all involved parties are compliant with it: all the corporation's technology, employees, both technical and end-users, and vendors that work with the corporation or provide solutions that are integrated in any way.

8.5 SECURITY USE CASES

Most of the critical infrastructures in our society depend on secure information exchange over communication channels. Some examples are energy management, financial transactions, and telecommunication networks in general. Secure information exchange is foundational to the critical infrastructures that underpin our society.

However, existing communication hardware has a limited lifespan and needs to be regularly updated, which makes it vulnerable to emerging threats. As quantum computing advances, it poses significant risks to the current cryptographic standards, potentially compromising the security of these critical systems.

Vendors are proactively enhancing their infrastructure to be quantum-safe with solutions for:

- Quantum links for long-distance communication in ground stations or space-based platforms.

- Quantum satellites to extend the network ranges on long distances with quantum communications.

- Quantum repeaters to extend the transmission distance by segmenting a long quantum link into shorter sections.

Providers must provide agile systems that are capable of regular updates. Such systems can incorporate new, quantum-resistant encryption solutions as they become available, ensuring broader applicability and enhanced security.

Quantum communication ensures security leveraging the no-cloning theorem, preventing any direct duplication of quantum information and eavesdropping, effectively co-propagating the classical and quantum signals at a long distance. It also demonstrates potential for increased information capacity, allowing more efficient and rapid data transmission. It provides resilience against emerging threats and will allow the creation of next-generation secure communication networks.

Some examples of telecom hardware vendors offering the new quantum-secure communications equipment are Apple, AWS, Ciena, Cisco, Hitachi, HP, Intel, Nokia, Toshiba.

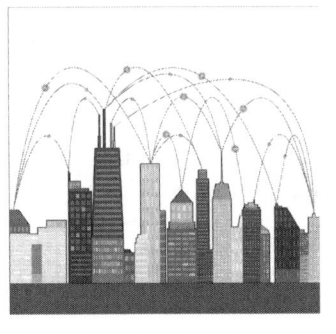	**Use Case: Secure communications**
	Business group: Telecommunications
	Problem Category: Communications
	Industry tests examples: BT, COLT, Orange, SK

Commercial networks need to securely transmit information of different content types, voice, text, or media, between parties over different distances. Classical data signal communication systems must co-exist with quantum ones.

Quantum communication provides resilience against emerging threats and will create next-generation secure communication networks.

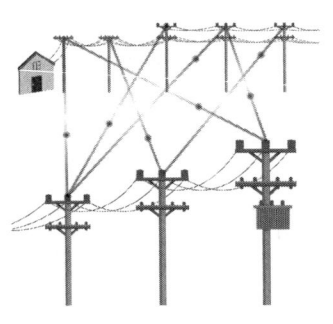	**Use Case: Secure Energy Transmission**
	Business group: Energy
	Problem Category: Communications
	Industry tests examples: Oman Grid, US DOE

The electric grid has a delivery control system to manage generation, transmission, and distribution of electricity to houses and it must communicate multiple energy devices over a single encrypted channel.

Encrypting communicated data within these systems is extremely important and should remain effective for the lifespan of the grid.

CHAPTER 9

Technology
Adoption Outlook

W E ARE CURRENTLY IN the early stages of the maturity of quantum technology and its adoption, with the different providers racing to produce good-quality qubits that can be isolated so that they do not interfere with each other and can maintain their properties. Tracking developments in quantum computing helps organizations plan, which is why it is important to understand how the Technology Adoption Cycle usually works. The figure below is a modification of Geoffrey Moore's Technology Adoption Cycle description to show how as qubit performance evolves, users start adopting the technology in different ways until it goes mainstream. It is mapped with an estimate of the time evolution, based on an analysis of main providers' qubit improvement rates.

As we can see, in the early stages, technology improvements are small and incremental, making them hard to notice on a day-to-day basis. However, these changes and tests accumulate over time, laying the groundwork for a significant shift. When the conditions are right, these accumulated changes will lead to a rapid and dramatic transformation, like it has happened with many other technology applications before, such as web pages, cloud computing, mobile apps, or artificial intelligence. In quantum computing, the improvement threshold will be achieved by reaching error correction.

Figure 9.1 shows the quantum technology evolution and adoption rate.

DOI: 10.1201/9781003302674-11

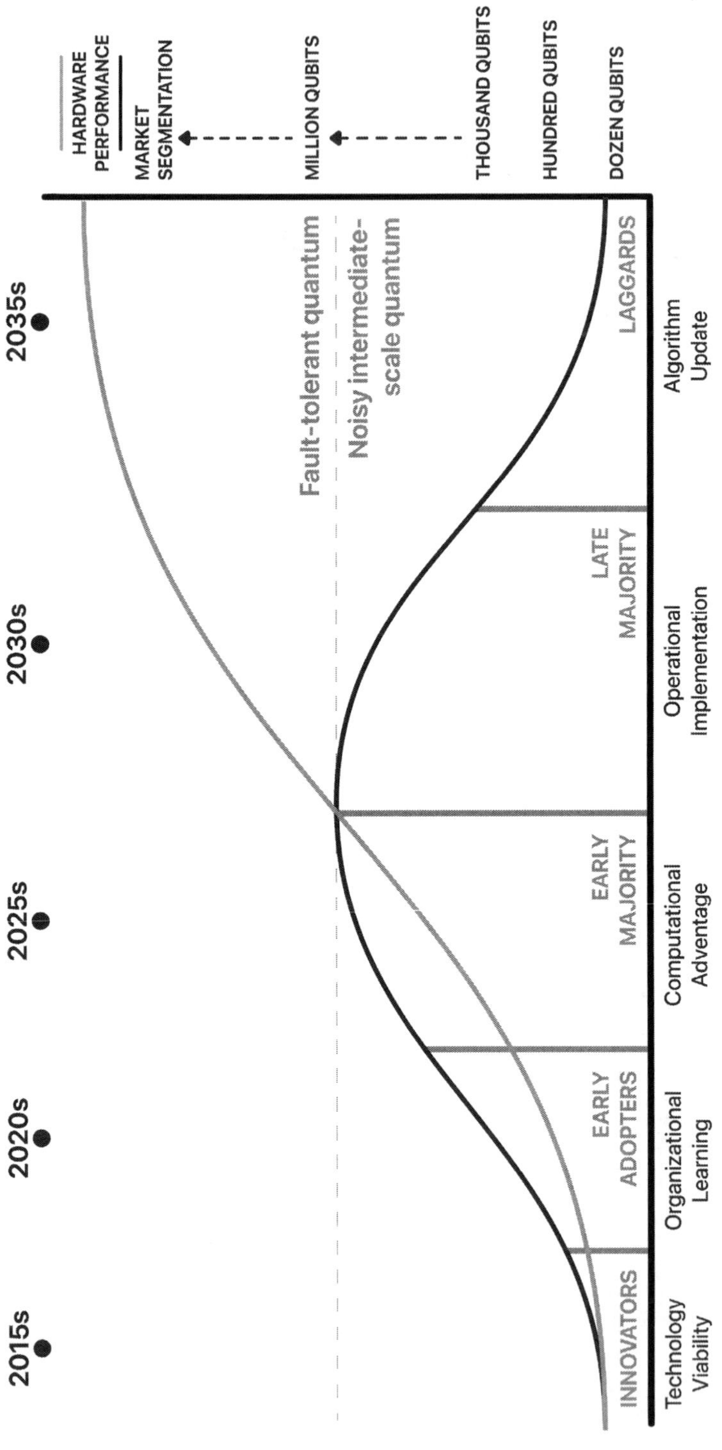

FIGURE 9.1   Quantum computing technology adoption.

At each stage of the technology adoption, the users have specific characteristics:

- **Innovators (2.5%):** Technology enthusiasts who want to be the first to try the technology; they are tech-savvy and have a high tolerance for bugs. They perform technology tests to check its viability and understand how it works.

- **Early adopters (13.5%):** Visionaries attracted by the high risk, they are part of the early market, and they test the technology with some support, willing to put up with some functioning errors. They develop conceptual use cases for organizational learning purposes, understand the potential technological advance, and know how to integrate quantum computing with their current systems.

- **The majority (68%)**

  1. pragmatists, the early majority, who want proven applications that work well without errors in a standard technology solution fashion, and

  2. conservatives, the late majority, who seek established solutions with practical business value. They test the technology usefulness and business competitiveness towards computational advantage and operational implementation.

- **The laggards (16%):** Those who will replace the technology when there is no choice, updating legacy algorithms as a last resort.

There is a need to have a better technology that will allow the development of solutions leading to a business advantage. As the technology evolves, companies start adopting it in different ways.

The **technology outlook**, as this book is being written, is at a point at which there are some real use case tests that show its potential advantage; we will see some of them in "Section II: Applicability" of this book. The advantage however is not at a maturity level that can bring real business value to corporations.

To understand the evolution stage of quantum computing technology, we consider the technology readiness level (TRL) metric developed by NASA in the 1970s to measure the maturity level of a technology as it progresses through the research, development, and deployment phases.

It is based on a scale of 1–9, with nine being the most mature. After all levels are completed, the solution is launched into the market. The TRL metric is helpful to figure out how well the technology works, what further development is needed to get it to properly work. It is also useful for benchmarking, risk management, and funding decisions.

Table 9.1 describes the generic TRL nine stages and Table 9.2 the Quantum Computing Technology Progression.

TABLE 9.1 TRL Definition	
TRL 1 – basic research	Scientific research is beginning to be applied. At this stage, a theoretical framework is formulated through research papers.
TRL 2 – applied research	The concept of the technology and its application are formulated through studies and experimentation.
TRL 3 – proof of concept	Studies and lab experiments conducted to assess technology viability, testing specific concepts and technology components.
TRL 4 – lab validation	Determines whether the solution works properly by integrating different technological components to create a pilot.
TRL 5 – industrial validation	Tests the pilot in a simulated setting to identify what needs to be improved to work in a real environment.
TRL 6 – system prototype	The technology is tested in a relevant environment to identify what needs to be improved to properly work.
TRL 7 – prototype demonstration	A solution is demonstrated to work in an operational environment without risks and a final design is completed.
TRL 8 – system complete	The system incorporates the commercial design after the tests have succeeded.
TRL 9 – system proven	The solution has its final form, it is certified and deployed for a commercial environment, ready for full-scale use.

TABLE 9.2 Quantum Computing Technology Progression	
Proof of concept ~ 2015–2025 [TRLs 1–4]	At this stage, the viability of the technology is tested. The tests performed mainly involve simple use cases, with the objective of understanding how they are affected by qubit noise and their requirements in terms of the gates needed. The aim is to improve algorithms with techniques to avoid noise propagation. This is done by experienced developers. Typically, a small number of qubits are used to perform analytical experiments to formulate the technical algorithms, and testing is performed with dozens to hundreds of qubits; validation is mainly conducted in a laboratory environment. *Quantum computers and applications are defined and tested.*
Realistic use cases ~ 2025–2035 [TRLs 5–7]	As the qubit quality increases, reducing existing noise in gates and operations, it is possible to obtain reliable and accurate solutions. It is now possible to validate the technology in relevant environments with easier-to-use developer coding tools and to create prototypes in operational environments to test both the use case performance and integration within a high-performance computing (HPC) environment. This is usually done with a larger number of qubits, on the order of thousands, with a lower error rate that allows multiple gates, that enable larger use cases to be tested in an operational environment. *Focused on error correction, needs and HPC integration.*

(*Continued*)

TABLE 9.2 (*Continued*)   Quantum Computing Technology Progression

Practical business ~ 2035 onwards [TRLs 8–9]	At this stage, we have entered a fault-tolerant era in which the qubit quality is good enough to enable a universal quantum computer.
	Coding tools are available for all developers that abstract quantum information concepts, making them easy to use. The potential of quantum computing has been fully demonstrated, and different use cases with specific business advantages are developed.
	This should require millions of fault-tolerant qubits, and such methods will be deployed in an operational production environment, that will also include post-quantum safe techniques to safely store the data.
	Computers are deployed, and business performance is tested.

SECTION II

Applicability

THE TERMS "INDUSTRY" AND "sector" are often used to describe a group of companies that operate in the same segment of the economy and have a similar business type. According to Investopedia, a sector is a large segment of the economy; studying sectors can provide insight into how an economy is performing and which areas of the economy are performing better than others. An industry is a more specific set of companies or businesses, consisting of a group of establishments engaged in the same, or similar, kinds of production activity.

There are four sectors: primary, secondary, tertiary, and quaternary.

DOI: 10.1201/9781003302674-12

The **primary sector** involves extracting and harvesting natural products from the earth, processing them, and packaging raw materials to sell them to consumers or businesses.
Activities include:
Mining and quarrying, fishing, agriculture, forestry, hunting.

The **secondary sector** produces goods from the natural products obtained by the primary sector, and it adds value to natural resources by transforming raw materials into valuable products.
Activities include automobile, textile, chemicals, aerospace, shipbuilding, energy utilities.

The **tertiary sector** concerns the production and exchange of goods manufactured by companies in the secondary sector; relevant industries include retail, entertainment, and finance.
Activities include retail sales, transportation and distribution, restaurants, tourism, insurance and banking, healthcare services, legal services.

The **quaternary sector** covers intellectual services involved in technological advancement and innovation as well as research and development (R&D) that lead to process improvements. Personnel in this sector work in offices, academia, hospitals, clinics, theaters, accounting, and brokerage.
Activities include R&D, information technology (IT), education, and consulting.

For the purpose of the analysis in this section, we combined industry and sector concepts to create business categories with similar characteristics to those defined by the OECD, Investopedia, National Geographic, and Encyclopedia Britannica:

1. Aerospace covers vehicular flight within and beyond the Earth's atmosphere.

2. Agriculture covers the preparation of plant and animal products for delivery or use.

3. Automotive covers the manufacturing of motor vehicles and their components.

4. Chemicals covers the processes and operations used to produce substances.

5. Energy produces or supplies resources such as fossil fuels and renewables.

6. Finance provides financial services to customers and keeps main street functioning.

7. Government provides a political system to administer and regulate a country.

8. Manufacturing covers the fabrication of components to yield finished products.

9. Pharma and healthcare provide medical services, equipment, and drugs for patients.

10. Supply chain manages the movement of goods from the manufacturer to the end user.

11. Telecommunications makes wired or wireless communication possible.

ANALYSIS METHODOLOGY

Each business category is analyzed in detail providing an overview of the industry/sector to contextualize them, following the framework described below.

Business Category Analysis Framework

 **Business category** is a short paragraph of the industry/sector, detailing its scope, economic impact, and the resources it relies on.

Explanation of the business group activities, its economic importance, employment rates, the main resources used, analysis of the value chain flow, and the main pain points.

Use Case Analysis Framework

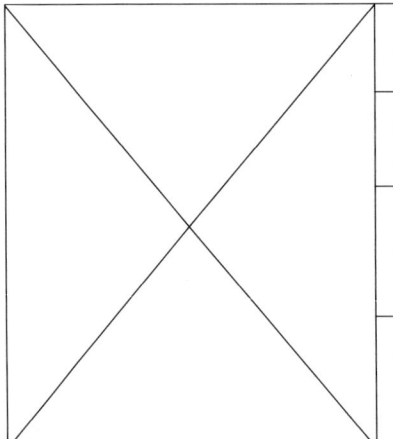

Short **use case** illustrative name.

Main **business group** the use case belongs to and can be more than one.

The main problem category to which the algorithms used to solve the use case belong, and there can be several.

Main **underlying technology** used to test the use case and has non-exhaustive list.

- Use case description of a specific situation to understand the background.
- Use case analysis to understand the existing problem and its impact.

Quantum computing value analysis includes benefits from a technical, business, or corporate point of view.

CHAPTER 10

Aerospace

The aerospace business covers aspects of mechanical air transportation and the activities that facilitate it.

Aerospace includes activities in the Earth's atmosphere and beyond, while aviation includes flying aircraft and operating planes.

Aviation accounts for 4.1% of the world's gross domestic product (GDP), and it is a powerhouse industry that generates more than 87.7 million jobs worldwide, as of 2022.

The main resources in aerospace include low weight materials, high strength, and resistance to heat, fatigue, cracks, corrosion, and loads, along with energy for flight.

This business group comprises the following activities:

- Manufacturing of airplanes, helicopters, gliders, lighter-than-air aircrafts, rockets (manned or unmanned).

- Aerospace systems, the design and use of the onboard systems found in aircraft or spacecraft.

- Production of spacecraft such as satellites and launch vehicles, either crewed or uncrewed.

DOI: 10.1201/9781003302674-13

The volatility of oil prices makes it difficult to predict fuel burn requirements, which can be damaging to operations in an environment where buyers demand high quality assurance from aircraft manufacturers. Aerospace provides the economic core to advance science and technology; it has high R&D investments and is capital intensive due to long-term projections of heavy investment in facilities or equipment that require long design lead times.

Figure 10.1 shows the aerospace activity value chain.

R&D and design
↓
Part & component manufacturing
↓
Subsystems/subassemblies
↓
Final assembly/integration
↓
Post-production services

FIGURE 10.1   Aerospace flow chart.

The business group's main pain points are related to the following:

1. Supply chain management challenges for production should be addressed; unique aircraft parts have long lead times that double production time from 25 to 50 weeks.

2. Environmental pollution should be reduced with efficient lightweight designs to balance time and fuel consumption, reduce emissions, and maintain aircraft.

3. The complexity and increase in the use of computer-based systems require higher quality assurance to avoid security problems and grounding of aircraft.

4. New energy sources should be found to decarbonize and transition towards an "all-electric aircraft".

5. There is a need to reduce the design time to cut investment costs, allowing for a higher flexibility to adapt quickly to the market needs and industry regulation needs.

Some companies testing quantum computing: Airbus, Boeing, Delta Airlines, DLR, Rolls-Royce.

10.1 USE CASE: ENGINE DESIGN

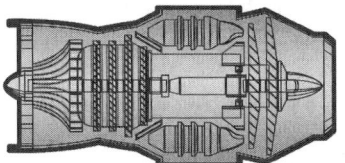

	Business Group: Aerospace
	Problem Category: Chemical Processes
	Underlying Technology: Simulation

The demand for new aero-engine products is growing and changing as alternative energy sources are introduced into the aircraft system that have new heat content, fluidity, corrosion protection, and stability conditions. The aircraft engine development is a comprehensive process that covers structural, mechanical, and systems engineering, as well as testing and verification towards an increased rate of performance while lowering harmful emissions and fuel burn.

To analyze jet engine performance, engineers use computer models to simulate aerodynamics and thermodynamics components and superheated airflow through them. These calculations are complex with a huge number of variables to test and can take days or even weeks to run, limiting the rate to test new designs. Engineers simplify and approximate models that can lead to inaccurate estimations.

Quantum computing enhances the efficiency of modeling air flow and hot gas in engine designs, significantly reducing the time required for simulations. This leads to the development of more efficient and environmentally friendly engines.

10.2 USE CASE: AIR CARGO LOAD

Business Group: Aerospace

Problem Category: Objective Optimization

Underlying Technology: Annealing | Trapped Ions

Airlines aim to make the best use of an aircraft's payload capacity as regulated by the US Federal Aviation Administration; this capacity is the maximum allowed takeoff weight minus the empty weight. The weight comprises all the aircraft equipment, minimum fuel load, oil, and flight crew. Ultimately, it affects fuel burn, overall operating costs, and therefore, revenue.

The scope for optimization can be limited by the operational constraints of the payload and the center of gravity, size, and shape of the fuselage. Optimization affects the efficient performance of an aircraft from the standpoint of altitude, maneuverability, rate of climb, and speed. As operations become more efficient, the overall number of required transportation flights could be reduced, with a positive impact on CO_2 emissions.

Quantum computing allows airlines to devise the most effective strategies for loading aircraft respecting flight constraints, leading to cost savings, enhanced aircraft performance, reduced environmental impact, and maximized plane capacity.

10.3 USE CASE: IRREGULAR OPERATIONS

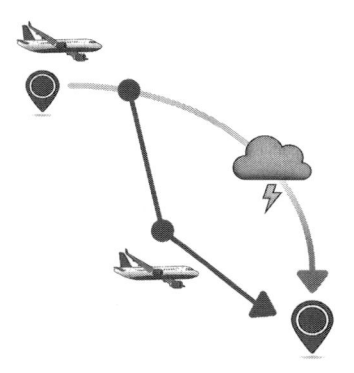

	Business Group: Aerospace
	Problem Category: Stochastic Simulation
	Underlying Technology: Superconducting

Airlines face extraordinary situations in which a flight does not operate as scheduled, defined as irregular operations management (IROPS), which can be due to weather conditions, technical glitches, or overbooking among others. The airline must decide if the flight should be cancelled, considering the network effect this will have in other flights as well as the cost, which is estimated to be 8% of an airline's revenue according to Amadeus research. They must also provide a solution to rebuild aircraft routings, reassigning crew, and rebooking passengers.

This problem is tackled in parts, focusing separately on finding solutions for the aircraft, crew and, lastly, passengers, because analyzing all the scenarios is computationally challenging due to the exponential growth of space size.

Quantum computing enables airlines to assess and simulate the potential impacts of various scenarios on their flight network more accurately. This broader analysis helps in minimizing disruptions, cutting unnecessary expenses, and boosting competitiveness.

10.4 USE CASE: FLIGHT CLIMB

	Business Group: Aerospace
	Problem Category: Artificial Intelligence
	Underlying Technology: Superconducting

The climb portion of a flight is when the engines operate at almost full power and maximum stress going through the dense atmosphere to reach the "en route" altitude at a thinner atmosphere. To do this, aircrafts must consider: speed, angle at which a plane climbs, flap settings, loading, weather conditions, and engine degradation, while keeping passenger comfort requirements of crawling time with specific fuel consumption standards.

Airplanes fuel consumption accounts for 25% of an airline OPEX or higher as prices increase. 10% of the fuel spent corresponds to the climb phase, since the fuselage and wings aren't "cutting" the air at their optimum angle while traveling at high speed with the engine full blast. A Boeing 747 uses about 14,400 L/hour, any time reduction in the climb phase decreases overall costs that can sum up to $50,000 per aircraft, per year.

Quantum computing, using genetic algorithms, can fine-tune the balance between speed and fuel consumption in aircraft, leading to both financial savings and a reduction in carbon emissions.

CHAPTER 11

Agriculture

The agriculture business group comprises ways people provide food and other goods by cultivating plants and breeding livestock to sustain the population's needs.

It is the backbone of the world's economy since it is the basis for food and nutrition and provides raw materials for industrialization.

It accounts for 4% of global GDP, it employs one-quarter of the labor force, and least developed countries rely on it as the primary income source, being more than 25% of its GDP.

The main resources for agriculture are sunlight, soil, and water, together with fertilizers.

This business group includes companies involved in producing food and crops for consumption with the following activities:

- Animal production segment for livestock.

- Production of plants cultivated in greenhouses, field crops, and food and grain crops.

- Farm management services.

- Floriculture, soil preparation, planting, and cultivation

DOI: 10.1201/9781003302674-14

Efficient operations are the primary profitability driver. Production, manufacturing, and distribution are heavily regulated, including the allowable residue levels in food. Before approval for sale, products are tested and evaluated to ensure they meet country regulations. There have been substantial increases in agricultural productivity costs attributed mainly to the growth in pesticides, which includes herbicides, insecticides, and fungicides. Pesticide use results in $1.5 billion in resistance, $1.4 billion in crop losses, $2.2 billion in loss of bird species, and $2.0 billion in groundwater contamination.

Figure 11.1 shows the agriculture activity flow chart.

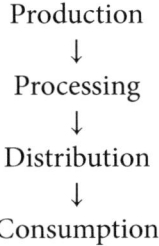

Production
↓
Processing
↓
Distribution
↓
Consumption

FIGURE 11.1   Agriculture flow chart.

Agriculture pain points are related mainly to pesticides and management:

1. Soil is becoming less usable, and crop production has declined by 40% in the past 50 years, while world population that needs to be fed has more than doubled.

2. There is a need for plant protection pesticides (PPPs) that take long to be produced, 11.3 years to develop, and cost $10 billion per year.

3. Median farm income and profit margins have steadily declined in the last 30 years with increasing production expenses. Soil conditioners and fertilizers account for $3.4 billion.

4. Soil pollution has become an environmental issue, agriculture is partly responsible for about 25% of greenhouse gas emissions.

5. Environmental changes and higher productions require increase in pesticide use and genetically engineered crops that cost an additional $1.1 billion.

Some companies testing quantum computing: BASF, Bayer, Center for Bioenergy Innovation, NASA.

11.1 USE CASE: EFFICIENT FERTILIZERS

| Business Group: Agriculture \| Government |
| Problem Category: Chemical Processes \| Objective Optimization |
| Underlying Technology: Superconducting |

Fertilizers are chemical substances that farmers use to improve crops, and almost all of them contain ammonia, the preferred nitrogen-containing nutrient for plant growth since it is essential to life. Tiny anaerobic bacteria in the roots of plants fix nitrogen naturally, turning it into fertilizer using the enzyme catalyst nitrogenases. The industrial Haber–Bosch process allows artificial synthesis of ammonia that needs high energy levels to operate, and the resulting soil fertilizer is easily soluble in water, which makes it easy to transport run-off waters.

The Haber–Bosch process accounts for 3% of natural gas burn and carbon emissions, relying on fossil fuels. Biological nitrogen fixation by the enzyme nitrogenases converts nitrogen into two ammonia molecules. This process is complex to engineer because of the vast number of possible catalysts and the reaction mechanisms to simulate molecule systems for nitrogen fixation. The fertilizer has a broad spectrum and about one-third is wasted, which can be solved by tailoring its production.

Quantum computers can accurately simulate nitrogen fixation analyzing molecular interactions which leads to sustainable and competitive alternatives to catalytic processes, and a reduced environmental footprint.

11.2 USE CASE: SEED BIOLOGY

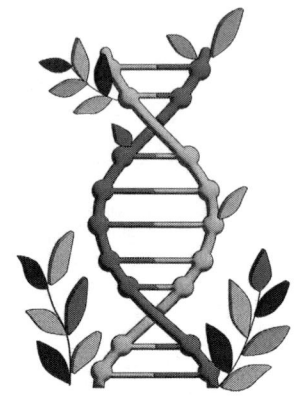

Business Group: Agriculture | Life Sciences
Problem Category: Stochastic Simulation
Underlying Technology: Superconducting

Seed quality is the basis of efficient crop production. Until recently, plant breeders' selection process was indirect since non-germinating seeds were naturally eliminated. With genomic analysis, regional variety tests are often conducted assessing molecular aspects of seed germination, to obtain agronomic information such as tolerance to disease and quality. This discipline complements traditional crop protection and allows to identify and select the best quality seeds with desirable traits, such as high yield potential or disease resistance.

With the increased know-how on molecular aspects, farmers can genetically modify seeds to improve their quality, and be more proactive to make them more resistant to threats. Today this is done in a trial and error process; this is a long process that can take on average 13 years, and it costs $130 million to bring a genetically modified seed into the market.

Quantum Computing can enhance genetic mapping and breeding accuracy, potentially increasing production yields, reducing water requirements, improving crop quality, and speeding up the time to market for new seeds.

11.3 USE CASE: WEATHER FORECAST

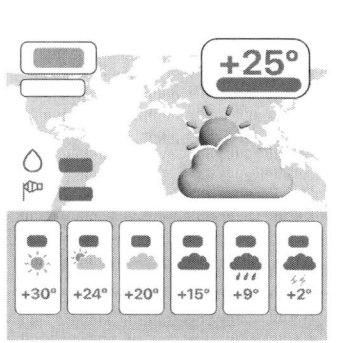

	Business Group: Agriculture | Aerospace | Energy | Government | Insurance
	Problem Category: Artificial Intelligence
	Underlying Technology: Neutral Atoms | Superconducting

The weather is a set of different meteorological events expressed in specific values at a particular point in space at a particular time. Its forecasting analyzes huge amounts of heterogeneous data containing several dynamic variables, such as air temperature, pressure, and density, that interact to predict the conditions of the atmosphere for a given location and time. Its goal is to protect human lives and property, and to improve health, safety, and economic prosperity.

Having accurate weather forecasts is only possible for short-term forecasts of less than a week, and long-term planning is crucial for large seed productions. It requires processing large amounts of data from various sources, such as satellites, ground sensors, and historical records. On top of this, the frequency and severity of extreme weather events have increased globally, with economic costs rising even more sharply. In agriculture, geographical accuracy is needed to control the crop in a specific location to be able to personalize it.

Quantum Computing can refine numerical methods to improve meteorological tracking, predict micro-meteorological events, and provide up-to-date information for operations sensitive to weather conditions, potentially increasing food production and well-being.

11.4 USE CASE: IMPROVED CROP YIELD

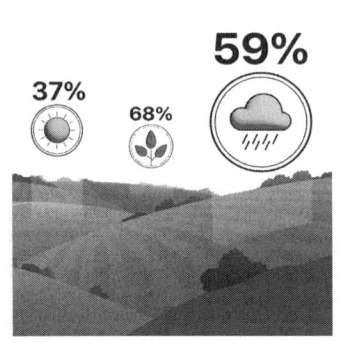

59% 37% 68%	Business Group: Agriculture | Government
	Problem Category: Objective Optimization
	Underlying Technology: Annealing

Crop production depends on the availability of arable land and is affected by yields, macroeconomic uncertainty, and consumption patterns. Balancing harvested areas with quantities produced is key to maintaining long-term sustainability of farms. Seed production planning to properly estimate what can be supplied, be able to produce, or advance, or delay harvesting as needed to have enough fresh regional food, without having to import it, has an economic impact.

Each type of plant has its own market and dynamics. The actual yield from a farm depends on several factors such as the crop's genetic potential, the amount of sunlight, water, nutrients absorbed by the crop, or the presence of weeds and pests. All these parameters affect each plant differently, depending on its type, and that will affect its production; for example, a simple increase in temperature decreases wheat production by 6%.

Quantum Computing can optimize harvest forecasts and the management of different crops in adjacent farmlands. This involves increasing the number of variables that can be processed simultaneously, leading to more efficient use of land, water, and energy resources.

CHAPTER 12

Automotive

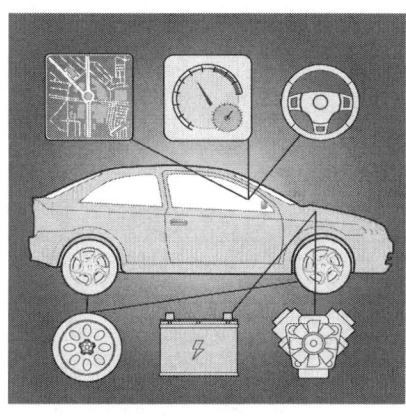

The automotive industry includes the manufacturing of motor vehicles used for transportation and their components, i.e., engines and bodies.

In 2021, the revenue of the global automotive manufacturing industry was $2.7 trillion.

It contributes to 27 million jobs globally and is one of the world's largest economic sectors by revenue. It also has the highest spending on R&D.

The primary products of the automotive industry are passenger automobiles and light trucks, including pickups, vans, and utility vehicles.

The business group consists of three key divisions:

- Motorcycles and cars, including passenger cars and light commercial vehicles.

- Trucks and buses designed to perform utilitarian work and transport items or people.

- Industrial-grade machinery for construction and agriculture.

DOI: 10.1201/9781003302674-15

Automobiles can be gas-powered, fuel, electric, or hybrid vehicles and can be autonomously driven. Companies can be categorized as manufacturing cars or parts of cars. This industry is a mixed oligopoly with a few leading global producers due to the expenses involved in manufacturing. This scenario is changing with new electric car regulations.

The industry relies heavily on capital investment for manufacturing facilities and distribution networks. It supports numerous employment systems, including mechanics, sales, assembly, finance, creative, scientific, technical, and business jobs.

Figure 12.1 shows the automotive activity flow chart.

FIGURE 12.1   Automotive flow chart.

The automotive industry's main pain points are related to the following:

1. Semiconductor shortages cost $210 billion in profits and a decrease in the production of vehicles by 7.7 million vehicles.

2. Reaching the target sales of zero-emission vehicles require the mass production of lithium-ion batteries or alternative materials.

3. Diagnostic software malfunctioning accounts for 40,000 deaths per year in the US.

4. With the change in industry dynamics, a 51% increase in the price of new vehicles, traditional original equipment manufacturers (OEMs) moving to services, and increased competition reduce the benefits.

5. Global distribution through automobile supply chains in accordance with local protocols decrease agile and robust inventory management.

Some companies testing quantum computing: BMW, Daimler Mercedes-Benz, Ford, Hyundai, Toyota, Volkswagen.

12.1 USE CASE: EV BATTERIES/FUEL CELLS

| Business Group: Automotive | Energy |
| Problem Category: Chemical Processes |
| Underlying Technology: Annealing \| Superconducting \| Trapped Ions \| Cold Atoms \| Photonics |

Rechargeable electric batteries power the motor of electric vehicles, they are made from a combination of raw materials, mostly lithium. To properly work, they must have a light weight, not to overheat, be able to last for long drive ranges, and have a long life. The goal is to maximize the electrochemical storage energy density capacity to extend the battery duration by testing electrode performance.

When designing batteries, balancing performance with safety is challenging. Analyzing the behavior of electrolyte particles and how they bind to each other can help to reduce battery costs up to 40%, improve the driving range to 1,000 km, and heat the car interior more efficiently. Unfortunately, computers are greatly limited in the number of electrons they can model to produce accurate predictions for new designs.

Quantum computing can precisely simulate the energy profiles of lithium batteries, enhancing the performance of electric batteries, lowering production costs, and contributing to sustainability.

12.2 USE CASE: TRANSPORT ROUTING FLOW

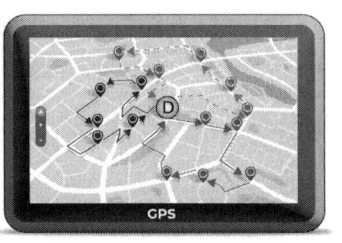

Business Group: Automotive | Energy | Government | Supply Chain
Problem Category: Objective Optimization
Underlying Technology: Annealing | Simulation

Vehicle routing aims to find the best possible route between a set of potential routes by minimizing the longest route length among all vehicles. This must be done while ensuring that each vehicle visits its required locations with constraints such as distribution time windows and vehicle load capacities. Solving this problem can reduce traffic congestion and air pollution and increase delivery productivity by anticipating and avoiding bottlenecks.

The problem becomes more difficult as time passes. The number of drivers and the vehicle's human-machine interactions provide increasing amounts of information, and its complexity makes it more complicated to analyze and explore all the solutions. Improving routing optimization can help to reduce operational costs and enhance the quality of delivery. It can save businesses up to 40% in driving time and fuel and reduce route planning time by 95%.

Quantum computing can determine the most efficient routes for vehicles in near-real-time, increasing operational efficiency, boosting profits, and decreasing fuel requirements, which benefits industries dependent on transportation.

12.3 USE CASE: OBJECT DETECTION

	Business Group: Automotive
	Problem Category: Artificial Intelligence
	Underlying Technology: Trapped Ions

To ensure the safe running of vehicles at high speed, real-time and accurate detection of objects such as road signs, pedestrians, animals, and cyclists is crucial. Autonomous vehicles use sensors and cameras to capture information about their surroundings and use computer vision techniques to detect visual objects of a certain class and identify what the object is and its location.

Detection still faces challenges when handling real-world objects under different viewpoints, illuminations, rotation, and scale changes and with dense or occluded objects. Algorithms are trained on massive benchmark datasets, and great effort is made to tune the hyperparameters to improve their predictive performance. The speed must be 0.033 seconds per image, and the accuracy must be close to 100% to be safe, neither of which has yet been achieved.

Quantum computing can optimize training algorithms to enhance accuracy, speeding up image classification and thereby making autonomous vehicles more dependable while also reducing their energy consumption.

12.4 USE CASE: AERODYNAMIC DESIGN

	Business Group: Aerospace | Automotive | Manufacturing
	Problem Category: Stochastic Simulation
	Underlying Technology: Cold Atoms | Superconducting | Simulators

Designing materials require a tradeoff between aerodynamic performance and structural weight while being able to withstand high forces. Simulations show the airflow distribution around an aircraft design using the Reynolds-averaged Navier Stokes nonlinear partial differential equations. This approach can be applied to the whole aircraft or a part. A change of less than 1% in the drag coefficient aerodynamics analysis can lead to failure.

Currently, it takes engineers years to model the air flowing process over a wing and associated turbulence. For some problems, the solutions are too complex to calculate, and most of the design is developed by making multiple variations and selecting the best via trial and error. However, an improvement to wing design can mean up to 30% fuel savings.

Quantum computing can accurately simulate the behavior of air atoms to study aerodynamic interactions, leading to faster design processes, improved performance, and reduced fuel consumption.

CHAPTER 13

Chemicals

Chemistry is the study of matter and energy and their interaction with each other. It enables the manufacture of a wide variety of products used in nearly every facet of human activity.

Some products are produced for direct purchase, but 70% of additives are used in other products.

Chemical businesses supported 120 million jobs globally in 2019.

Chemicals businesses generated $5.7 trillion of the global GDP in 2019, its main resources used in manufacturing chemicals range from air to minerals to oil.

This business group is divided into three main sectors: basic, specialty, and consumer chemicals, and comprises the following activities:

- The basic chemicals used for production include petrochemicals, chemicals derived from oil, polymers, and basic inorganics.

- Specialty chemicals include a diverse range of chemicals produced for agriculture.

- The consumer chemical sector includes toiletries, cleaning supplies, and cosmetics.

DOI: 10.1201/9781003302674-16

The rising costs of raw materials and transportation continue to increase operation expenses and limit the availability of high-demand products in a volatile market.

This industry is prone to rapid changes in consumer demand for new products and continued efforts to access raw materials and energy, improve processes, and impact R&D strategies.

Figure 13.1 shows the chemicals activity flow chart.

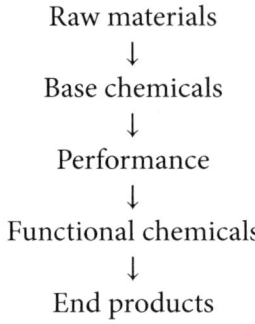

Raw materials
↓
Base chemicals
↓
Performance
↓
Functional chemicals
↓
End products

FIGURE 13.1   Chemicals flow chart.

The chemicals business group pain points include the following:

1. The risk of hazardous exposure to chemicals during transportation costs chemical plants $170 billion annually to maintain workplace safety, in addition to $250 billion each year directly associated with workplace fatalities and injuries.

2. The fluctuating demand for raw materials causes spikes in demand, making timely delivery difficult.

3. Inventories must be managed efficiently to handle market trends that increase the need for chemicals.

4. An increase in the cost of oil affects both the transportation and the cost of raw materials, which are scarce.

5. Meeting environmental demands to reduce carbon footprints and transition to more sustainable chemical products, including catalysts and surfactants, can contribute to the reduction of 100 million tons of CO_2 and costs over $500 billion annually.

Some companies testing quantum computing: BASF, BP, BMW, Dow, Johnson Matthey, LG Electronics, Mitsubishi Electric.

13.1 USE CASE: UNDERSTANDING MOLECULAR PROPERTIES

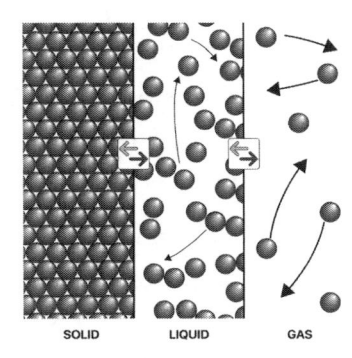

	Business Group: Chemistry
	Problem Category: Chemical Processes
	Underlying Technology: Superconducting

SOLID LIQUID GAS

Understanding molecules' properties, both chemical and physical, is crucial to use them for research or product development. The molecular energy states of a system of interacting electrons are calculated using the Schrödinger equation to analyze energy differences, gradients, and electrostatic properties. There are many possible combinations of atoms, and many possible ways that they can bond when developing new molecules.

The status of a particular electron in a molecule, especially during a chemical reaction, has many possible values. Describing the location and movement of electrons take an enormous amount of processing power and memory. The main challenge to analyze electronic structure is the computational cost of calculating the many electron wavefunctions, since the problem has combinatorial growth with the size of the studied system.

Quantum computers can precisely model correlation energies, improving quantum chemical calculations for a deeper understanding of molecules. This leads to a reduction in the need for trial and error tests.

13.2 USE CASE: DESIGN OF AGGREGATES

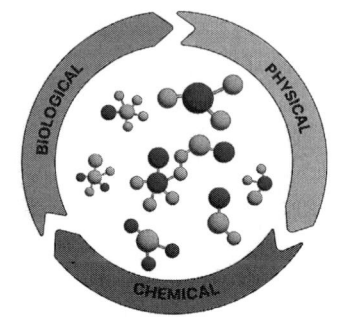

Business Group: Chemistry	
Problem Category: Artificial Intelligence | Objective Optimization	
Underlying Technology: Cold Atoms	

Understanding material compounds involves accurately describing the characteristics of all their molecules, the particles that constitute them, and simulating how they react under various circumstances. The design of new small molecules or polymers relies on accurate predictions of molecular properties to select which ones are the best, which is achieved by determining the probability distribution of electrons in a molecular system.

Solving the electronic structure of all molecules to find the best properties is a large task, since there are many desirable properties to test and many configurations that may fit. Because of the computational limitations we can obtain "exact solutions" only for individual atoms and very small molecules; the rest must be an approximation.

Quantum computers can efficiently optimize atomic composition and electronic structures in a vast array of molecules, enhancing material design, opening new revenue streams, and promoting sustainability.

13.3 USE CASE: CRYSTAL STRUCTURE

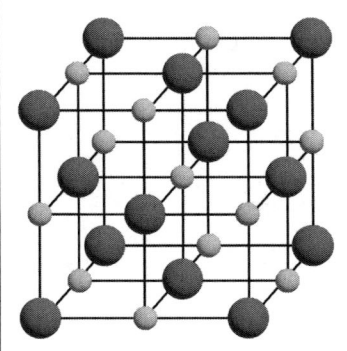

	Business Group: Chemistry
	Problem Category: Objective Optimization
	Underlying Technology: Trapped Ions| Photonics| Adiabatic

The analysis of a molecule involves analyzing its geometrical shape to understand how the atoms are arranged and where they are located, and to determine the bond lengths and angles. The molecular structure is critical to determine the physical, chemical, and biological behaviors that affect the properties of a chemical compound. The molecular structure of a compound can help determine its polarity, reactivity, phase of matter, color, magnetism, and biological activity.

Identifying a crystal structure involves analyzing the arrangement of the elements in a different lattice in the molecules. To achieve this task, we make predictions and observe the lattice to check if the predictions are correct. This is a computationally challenging problem as we must simulate all the possible structures until we find the specimen with the right one. The number of different configurations grows exponentially, and obtaining high-quality crystals is difficult.

Quantum computing enables the accurate prediction of molecular energies, facilitating faster product development that can lead to increased revenue, improve competitiveness, and decrease environmental impact.

13.4 USE CASE: CHEMICAL REACTIONS CATALYSTS

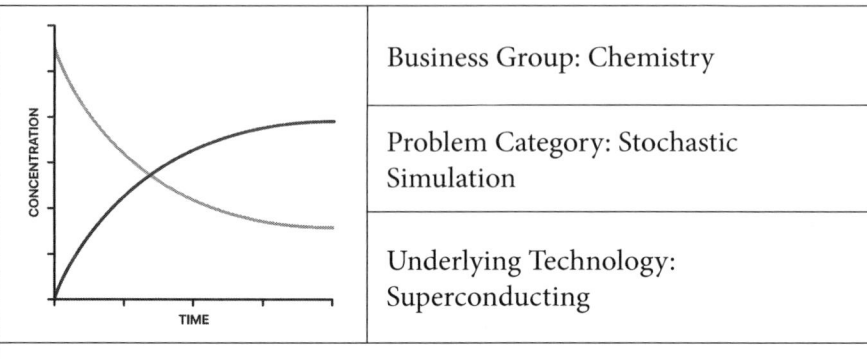

	Business Group: Chemistry
	Problem Category: Stochastic Simulation
	Underlying Technology: Superconducting

A chemical reaction is a transition from one molecular arrangement to another. The reaction path describes the various steps or stages along this journey, analyzing the molecules, the reactants, individually and in combination, to form the product. We must understand the thermodynamics, the electronic structures, and at what speed a chemical reaction occurs. We also need to predict different reactions and how they occur.

We use catalysis to drive reactions by reducing energy barriers, so we use a catalyst (another molecule) that can activate the molecules by reacting. This process uses the same reactants with a lower transition state to obtain a product. We must understand the molecules, how to activate them, and how they change. These tests can be dangerous to conduct, and they can be expensive, computationally demanding, and time-consuming.

Quantum computers can simulate the interactions between catalysts and reactants to encode and analyze many individual interactions simultaneously to improve the discovery and optimization of catalysts, surfactants, and solvents.

CHAPTER 14

Energy

The energy business covers the production and sale of energy, from fuel extraction, manufacturing, and refining to distribution.

Companies are categorized based on how the energy they produce is sourced: non-renewable or renewable, with electricity as a secondary source.

The oil and gas industries employ nearly 12 million workers worldwide.

The energy market is led by the oil and gas industries, with more than 50% of the worldwide energy demand. Approximately 30 billion barrels of oil are consumed yearly, primarily by developed countries.

The business group has four different types of companies, each with a role in bringing energy to clients. It consists of the following activities:

- Oil and gas drilling and production companies: they perform upstream activities to drill, pump, and produce oil and natural gas; pipeline and refining midstream activities to transport oil and gas from production sites; and downstream activities to convert crude oil and raw natural gas into final products.

- Renewable energy companies: they harness power from water, wind, or the sun to avoid 'lock-in' of CO_2 emissions.

DOI: 10.1201/9781003302674-17

- Electricity production and distribution companies: they are powered by a mix of coal, natural gas, nuclear, and renewables such as water and solar.

- Mining companies: they are coal companies that power plants, including nuclear plants.

Energy businesses are highly sensitive to macroeconomic factors and government regulation.

Figure 14.1 shows the energy activity flow chart.

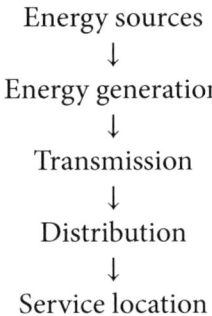

FIGURE 14.1   Energy flow.

The energy business group's main pain points are related to the following:

1. Preproduction capital is high and tied to 10- to 20-year investment cycles for deposits located during the exploration period.

2. The supply chain is complex and inefficient, with many raw materials coming from distant parts of the world.

3. Supplies are abundant but are non-renewable resources, posing environmental challenges with pressure to supply cleaner energy and reduce emissions.

4. A distributed control grid with a growing mix of energy sources creates the need for a smart grid to choose the appropriate mix to meet demand.

5. An increase in carbon capture and storage to deploy renewable energy and support new technologies requires a $46 trillion investment.

Some companies testing quantum computing: Eni, EDF, E.ON, ExxonMobil, RES independent renewable energy, Rolls-Royce.

14.1 USE CASE: RESERVOIR SIMULATION

| Business Group: Energy |
| Problem Category: Chemical Processes |
| Underlying Technology: Cold Atoms |

Reservoir simulation is a process used to simulate and analyze the behavior of hydrocarbon reservoirs through mathematical models that represent the subsurface reservoir to predict its fluid flow dynamics in porous materials, encompassing oil, water, and gas. By predicting the geological characteristics of the reservoir and the area surrounding it, it is possible to find the best location for energy plants, efficiently drill, extract the hydrocarbon, and optimize its transportation from underground to the surface.

Fluids, such as oil, gas, or water, can blend with the ground rocks; obtaining only the chemical we need and avoiding waste can be challenging. The process to extract them uses several methods, such as a transmitter-receiver system that obtains samples from the ground to analyze it and define the strategy or hydraulic fracturing to inject fluids to fracture the rocks and release fluids. This method is inefficient, creating an extra cost of up to 30%.

Quantum computing can simulate complex molecular interactions within subsurface geology, offering insights for efficient drilling decisions. This aids in predicting reservoir performance, enhancing hydrocarbon yield, and minimizing waste.

14.2 USE CASE: ENERGY UNIT COMMITMENT

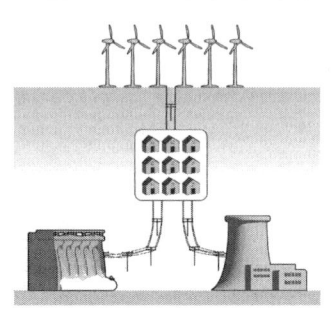

	Business Group: Energy
	Problem Category: Objective Optimization
	Underlying Technology: Annealing

The objective of the unit commitment problem (UCP) consists in minimizing the cost of power production, for a certain time horizon, having the best layout and schedule to generate and serve energy, which includes the operational costs associated with fuel consumption, maintenance, and start-up or shutdown operations. Solving the problem helps power system operators and utilities choose which units to commit and dispatch to ensure reliable and cost-effective power supply.

Solving the problem requires sophisticated optimization algorithms that consider the complexity of the power system and the constraints involved. As the power system size and complexity increase, finding the optimal solution becomes computationally challenging since the time required to solve the UCP grows exponentially with the number of power units and time intervals considered. Additionally, power systems require real-time decision making to maintain a balance between supply and demand.

Quantum computing can refine wind farm layouts to harness more energy using the same resources, thus decreasing costs, and improving efficiency and reliability. This also facilitates the adoption of new energy sources.

14.3 USE CASE: SMART-GRID OPERATION

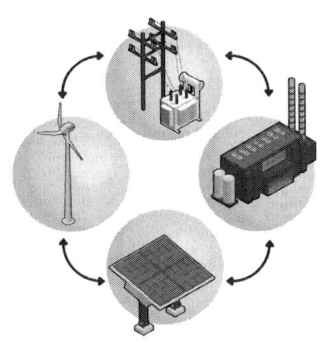

Business Group: Energy | Telecommunications | Government	
Problem Category: Artificial Intelligence	
Underlying Technology: Photonics | Superconducting	

With the rapid proliferation of renewable energy sources such as solar, wind and hydro, and even households as generating units, energy grids must be able to predict demand and production. To do this, they need to identify the best electricity-generating unit to meet electric power demand while balancing the usage cost with resource availability in real time. This goal is achieved through an advanced electricity distribution network that enables bidirectional communication between all involved parties, from any source, and data analysis for energy management.

Renewable energy sources lack a consistent base load. They depend on the inherent intermittent availability of their natural sources, which is affected by weather conditions and other factors. This leads to fluctuations in power outputs and added complexity in managing the activation and deactivation of transmission lines, making it difficult to respond to sudden changes in customer demand, power outages, sudden drops and rises in renewable energy output, and any other events.

Quantum computing is capable of processing diverse datasets to forecast and optimize energy demand and management across various systems. This reduces operational costs and integrates renewable energy sources.

14.4 USE CASE: GAS TURBINE DESIGN

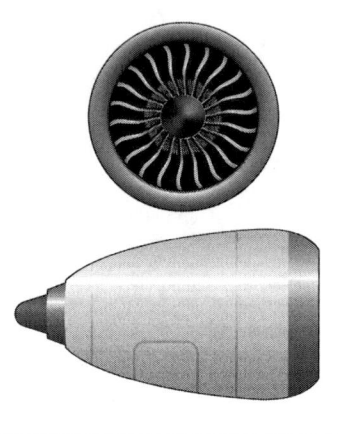

	Business Group: Energy | Aerospace | Manufacturing
	Problem Category: Stochastic Simulation
	Underlying Technology: Photonics

With the rise of alternative energy production methods, it has become vital to analyze the fluid and air flow characteristics to optimize the design, position, and placement of energy generation devices. Turbines are exposed to variations in environmental conditions, such as humidity, temperature, and salinity, which can change over time and affect their performance. Fluid and air flow rates, turbulence, and geographic positioning impact the density level of the energy source.

Optimizing energy generation device configurations to adapt turbines to environmental factors that can break or corrode them can extend their lifetime by up to 25%. Because individual turbine positions are highly correlated, the turbine positioning and layout can be adjusted to minimize the wake effects caused from one turbine to another and to maximize the expected power production. However, analyzing the number of possible configurations is in the order of millions and hence complicated to calculate.

Quantum computing can expedite the design process, optimizing the balance between the inner aerodynamics of a turbine and its overall weight. This enhances their efficiency and increase their lifespan.

CHAPTER 15

Finance

According to the IMF, "financial services are the processes by which consumers or businesses acquire financial goods".

Financial services is the primary driver of a nation's economy, as they enable the free flow of capital and liquidity in the marketplace.

It makes up approximately 25% of the world economy, with a market capitalization of approximately $22.5 trillion in 2021.

Financial services' main resource is capital, which is supplied by deposits, loans, securities, and funds.

This business group covers three specific categories: banking, investing, and insurance and it comprises the following activities:

- Banking focuses on loans, deposits, and payment handling, generating revenue through net interest and fees.

- Financial Markets center on investments and trading in stocks, bonds, forex, and derivatives, aiming for future gains.

- Insurance offers financial loss prevention and risk management for events such as health, property, and life-related risks.

DOI: 10.1201/9781003302674-18

The finance industry's strict regulations and risk mitigation demands have led to a 20% increase in operational costs over the past decade.

Figure 15.1 shows the financial services activity flow chart.

Product development
↓
Product distribution
↓
Customer acquisition
↓
Customer relationship management
↓
Transaction processing and operations
↓
Risk management and compliance

FIGURE 15.1   Financial services flow.

The finance business group's main pain points are related to the following:

1. Noncompliance fines related to know-your-customer (KYC) process can reach €3.5 million.

2. Inefficiency leads to 56% of potential clients leaving onboarding and raises churn rates by up to 25%.

3. 40% of decisions are data-driven, benefiting tech giants like Amazon, Google, and Meta, which leverage advanced analytics and now lend over $1 billion annually.

4. Accurate risk management and capital estimates are key to meet Basel III standards.

5. With shifting business models and growing demand for digitalization, 61% of finance companies expect digital, personalized interactions that could triple revenues.

Some companies testing quantum computing: Banco Santander, Bank of Montreal, BBVA, BMO, Citi, Credit Mutuel, Erste group, Financial Group, Bradesco, CaixaBank, Goldman Sachs, HSBC, Itaú, JP Morgan, Nomura, PayPal, Scotiabank, Truist Bank.

15.1 USE CASE: PORTFOLIO MANAGEMENT

	Business Group: Finance
	Problem Category: Objective Optimization
	Underlying Technology: Annealing| Superconducting

Investment portfolio optimization consists of choosing assets that, when combined, can help customers earn greater returns based on factors such as investor and risk profiles. Portfolio management involves computationally intense models that process large sets of variables.

However, as assets are added to a portfolio, the complexity of managing all the investment option variables to simulate all scenarios and validate risk sensitivities under market changes becomes excessive. Properly rebalancing portfolios to align with the market while handling all associated fees (taxes, commissions, etc.) can save up to 50% of the transaction costs and yield $600,000 savings in trading costs for a four-asset $1 billion portfolio.

Quantum computing can accelerate the optimization of investment portfolios through more effective rebalancing. This enhances revenue generation, lowers associated fees, and offers a strategic edge to financial institutions.

15.2 USE CASE: FRAUDULENT TRANSACTIONS

Business Group: Finance | Aerospace | Telecommunications

Problem Category: Artificial Intelligence

Underlying Technology: Superconducting

Companies search large amounts of constantly changing data to detect fraudulent activity in tax forms, transfers, or credit cards to make customer creditworthiness decisions. The datasets to analyze are large, with millions of samples with up to 10,000 different properties, including IP addresses and device types or locations.

The process of feature selection is a critical step in building powerful models to fight sophisticated fraud. Finding patterns in data has become excessively complex for current unsupervised deep learning models, resulting in data misclassification. As a result, fraud detection systems remain highly inaccurate, returning 80% false positives, which costs up to $40 billion in revenue loss and damages corporate reputations.

Quantum computing can refine fraud detection models, identifying hidden patterns, increasing their accuracy and efficiency. This reduces the resources spent on investigating false positives improving brand perception.

15.3 USE CASE: PRODUCT PRICING ACCURACY

	Business Group: Finance
	Problem Category: Stochastic Simulation
	Underlying Technology: Photonics| Superconducting| Trapped Ions

Options are derivative contracts that give the buyer the right to buy or sell a set amount of the underlying asset at a strike price before the contract expires. Monte Carlo methods, commonly used to estimate prices, make use of a large set of variables, including the underlying stock price, exercise price, volatility, interest rate, and time to expiration.

An estimate of the probability that an option will be exercised is usually obtained by randomly simulating different asset price paths to see how they can change over time and converge to an answer. Generally, 1,000 or 10,000 runs are recommended for calculations instead of determining the optimal number of runs needed. There is a lack of precision of approximately 1%, causing mistakes in VaR computing, hedging, and accounting.

Quantum computing can decrease the number of samples required for simulations, thereby streamlining the process. This improvement leads to greater efficiency, enhanced revenue, and heightened competitiveness.

CHAPTER 16

Government

A government regulates and operates countries, with responsibilities over R&D to keep the nation competitive and innovative, while ensuring national safety.

The scope of work includes the national, state, and local governments and public authorities.

As an example, in 2023, US federal revenue was 16% of GDP, generating 2 million jobs, with defense spending taking nearly half of discretionary spending.

Government functions are split into distinct institutions constituting branches with particular powers, legislature, executive and judiciary.

This business group, usually comprises the following activities, among others:

- Defense and Military Departments provide forces to ensure security.

- Energy department to address energy, environmental and nuclear challenges.

- Transportation is responsible for planning and coordinating all transportation means.

- Health Care improves citizens' well-being through medical advancements, and public health support.

- Communications ensures connectivity and manages spectrum use.

DOI: 10.1201/9781003302674-19

Governments strive for economic growth to enhance living standards by increasing production and improving nutrition, housing, safety, and healthcare. They fund and manage organizations that provide essential public services, including education, healthcare, public security, and transportation.

Figure 16.1 shows the government activity flow chart.

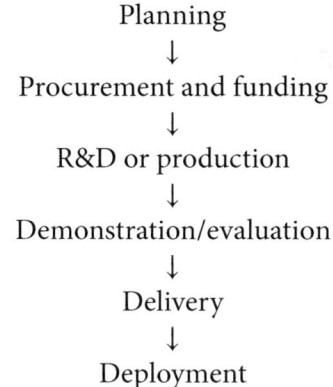

Planning
↓
Procurement and funding
↓
R&D or production
↓
Demonstration/evaluation
↓
Delivery
↓
Deployment

FIGURE 16.1   Government flow.

The government business group's main pain points are related to the following:

1. The number of cyberattacks reached 304 million in 2020, and governments have proactively prepared responses.

2. Nearly 50% of government systems are outdated, lacking digitalization and causing inefficient operations, slowing modernization progress.

3. Inefficiency costs $78 billion annually due to unclear structures and roles, failing to meet citizens' needs.

4. The fight against climate change continues, with extreme weather events like hurricanes and floods causing up to $3 billion in damages per event.

5. Strict regulations and bureaucracy add $125 billion in unnecessary administrative costs.

Government agencies and companies testing quantum computing: Department of Energy, DLR, NASA, National Energy Technology Laboratory, Thales, Total Energy, US Navy and Air Force.

16.1 USE CASE: CARBON CAPTURE SUSTAINABILITY

Business Group: Government	Energy	Chemicals
Problem Category: Chemical Processes
Underlying Technology: Superconducting

Carbon capture, utilization, and storage (CCUS) is the process of capturing carbon dioxide emissions prior to release into the atmosphere to permanently store carbon dioxide safely in deep geologic formations, use the carbon to enhance energy production, or use the oxides to make materials to help reduce the carbon intensity of industrial operations. The process requires equipment to separate the carbon oxides and transport them.

The CCUS process is complex, expensive, and energy intensive, requiring new power plants, which contribute to pollution creation. The goal is to accelerate the capture process and increase the CO_2 ready for transport or use. Chemicals such as liquids, solids, or membrane solvents are used for production, and higher compression techniques enable improved transportation mechanisms and geographic storage options in different geographic locations. Creating CO_2 with higher partial pressure can reduce production costs by 60%.

Quantum computing can advance the development of more efficient solvents, enhancing carbon capture and storage mechanisms. This innovation can reduce industrial costs, increase CO_2 sequestration, and aid in combating climate change.

16.2 USE CASE: TRANSPORT EFFICIENCY

Business Group: Government |
Aerospace

Problem Category: Objective
Optimization

Underlying Technology: Annealing |
Superconducting

Transport signaling for air and land is a complex task, but improving routes can minimize traffic jams. The task includes route identification, speed specifications, and conflict resolution techniques. An optimal planning scenario calculates multiple routes or operating approaches for each transport, sorting by cost (e.g., fuel, payload, congestion), and choosing the one that best fits the objectives.

Dynamic route optimization combines physics data on wind, temperature, and airplane performance, with restrictions imposed by traffic control associations such as flight permissions or company itinerary policies. The calculation size and complexity can reach hundreds of thousands of individual calculations for a single flight. Using fixed route data costs an extra 1 million gallons of fuel and 20 million pounds of CO_2 emissions annually.

Quantum computing can simultaneously analyze different transport modes and conditions boosting operational efficiency for customer service. This leads to reduced fuel consumption and improves financial performance.

16.3 USE CASE: SATELLITE IMAGING

	Business Group: Government
	Problem Category: Artificial Intelligence
	Underlying Technology: Photonics | Cold Atoms

Satellite mission planning plays a key role in earth observation, imaging for meteorology, surveying, and mapping imaging. Remote sensing planning is performed according to the satellite system elements, ground stations, observation targets, and the environment. Missions must be optimized in terms of the number of satellites to employ them efficiently, since they are difficult to manufacture, expensive to maintain, and challenging to launch into orbit.

Estimating the factors affecting the missions and planning satellites accordingly is a complicated task since they can range from simple, single-purpose satellites to complex, multipurpose systems with numerous subsystems and components, which together with the mission's objectives impact the cost. Maximizing the number of high-priority tasks completed on real datasets containing thousands of tasks and multiple satellites is computationally challenging. In large-scale missions, the size of the problem requires a prohibitive number of computational resources.

Quantum computing can optimize the scheduling of space missions, enhancing the efficiency of earth satellite operations. This improvement supports global monitoring of various earth activities and carbon footprints.

16.4 USE CASE: MILITARY OPERATIONS

	Business Group: Government
	Problem Category: Stochastic Simulation
	Underlying Technology: Superconducting

To devise battle strategies, military personnel evaluate numerous alternative courses of action for their own forces and the opposing side to help formulate operational decisions while effectively mitigating the positive or negative impacts of uncertainty in evaluating the risks. These strategies combine aspects of continuous change with intent, decide a course of action, calculate battlefield movements, and estimate probabilities for state transitions, both to prune truly unlikely possibilities and to help track execution.

The number of possible futures can be very large due to the number of entities involved, and the likelihood of those futures must be calculated, usually by means of Monte Carlo simulation to generate samples from the space of possible outcomes. These calculations are computationally costly without any guarantee that the space of behaviors is sampled adequately. Some challenges faced are the many unknown parameters, the need to efficiently handle multiple objectives and multiple constraints, and the curse of dimensionality for multiple criteria.

Quantum computing can refine simulations to better handle complex scenarios. This leads to improved strategic planning, reducing the need for expensive physical war-gaming exercises, and enhancing national security.

CHAPTER 17

Healthcare Life Sciences

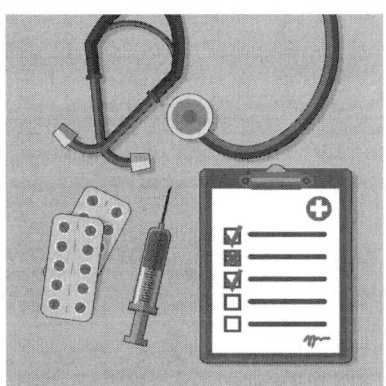

Healthcare life sciences encompass businesses providing medical services, insurance, medical equipment, pharmaceuticals, and patient care. Treatments involve small molecules, biologics, nucleic acids, blood products, vaccines, and peptides.

The sector spans four key categories, overseeing research, manufacturing, and facility management. With healthcare R&D comprising 20.8% of global business R&D spending, the pharmaceutical industry is valued at $1.42 trillion worldwide.

This business group comprises the following activities:

- Drug manufacturers include biotech firms for R&D, pharmaceutical firms for producing and marketing drugs, and generic drug manufacturers.

- Makers of medical equipment produce standard medical apparatuses, supplies, or hi-tech devices.

- Managed healthcare companies provide health insurance through individual firms.

DOI: 10.1201/9781003302674-20

- Facility management firms run hospitals, clinics, labs, psychiatric centers, and nursing homes.

The industry has high government regulation, with entry barriers from licensing, IP protection, specialization, and R&D costs.

Figure 17.1 shows the healthcare activity flow chart.

Drug Discovery: target proposal
↓
Drug Discovery: molecule search
↓
Preclinical research
↓
Clinical development
↓
Market approval

FIGURE 17.1   Healthcare Life Sciences (Pharma Focused) flow chart.

The pharma business group's main pain points are related to the following:

1. Administrative costs, driven by regulation and insurance complexities, account for nearly one-third of US healthcare spending.

2. The shift from chemical agents to alternative methods increases medicine manufacturing costs.

3. Diagnostic exploratory tests and scans are extremely expensive.

4. Chronic conditions contribute to more than 75% of healthcare costs.

5. Patients lack control over supply costs or equipment choices, with supply chain management being a major healthcare expense.

There is a high demand for healthcare due to the aging population and population of baby boomers.

Some companies testing quantum computing are Biogen, Boehringer Ingelheim, GSK, Merk & Co, Moderna, Pfizer, Roche.

17.1 USE CASE: DRUG CANDIDATES

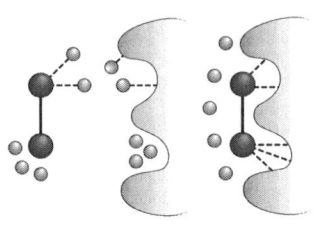

	Business Group: Healthcare & Pharma
	Problem Category: Chemistry Processes
	Underlying Technology: Annealing| Superconducting

Millions of potential molecules are tested through computer simulations (in silico) to find which will be the best compound combination to treat a pathology, reducing the number of potential molecules to test later in vitro. This is done by testing binding affinity, which reflects the strength of the interaction between the drug and its target by gaining insights into the molecular mechanisms of protein-ligand interactions by molecular docking and molecular dynamics simulation.

Computer-aided drug discovery for the simulation of molecular interactions to analyze the binding strength must be accurate since this corresponds to the needed local drug concentration at the target, determining drug efficacy, which translates into the projected therapeutic human dose. The accuracy and speed of these methods are limited, mainly due to simplifications made throughout the process, including neglecting the presence of water, the system's dynamic nature, and flexible interactions between the protein and the ligand.

Quantum computing can model the intricate biological activities of molecules, facilitating more efficient synthesis and decreasing the need for experimental trials, thereby reducing overall research costs.

17.2 USE CASE: MEDICAL IMAGING

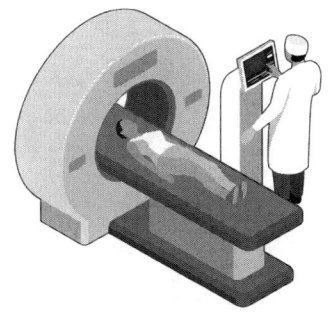

	Business Group: Healthcare & Pharma
	Problem Category: Artificial Intelligence
	Underlying Technology: Annealing| Superconducting

Medical images enable doctors to see inside the patients' body for diagnosis purposes. Medical imaging involves processing 3D image datasets of the human body obtained from different imaging device modalities, such as X-ray, computerized tomography, magnetic resonance imaging ultrasound, and positron emission tomography scan, each unique in terms of the images it gathers, equipment it uses, and what can be analyzed. The imaging process consists of acquiring raw data from the imaging devices and reconstructing the data into a format suitable for use in relevant software.

Imaging collects and interprets multiple sources of information over time considering each element's context. It presents enormous data processing challenges that lead to diagnostic errors in medical imaging. Average diagnostic error rate estimates range from 3% to 5%; there are approximately 40 million diagnostic errors involving imaging annually worldwide. In diagnostic imaging, raw information is released in pixel format, making it hard to obtain a high-level representation of tissues, organs, and lesions, to combine with clinical annotations of the region of interest for analysis.

Quantum computing has the potential to enhance medical image classification, allowing for more nuanced contextual analysis that improves disease diagnosis and patient treatment, reducing the likelihood of errors.

17.3 USE CASE: PROTEIN PATHOLOGY

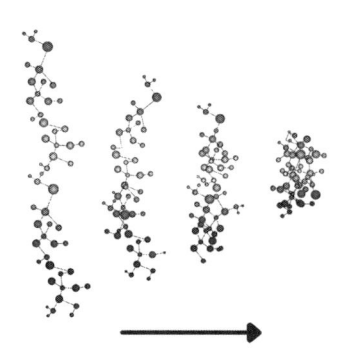

	Business Group: Healthcare & Pharma
	Problem Category: Chemical Processes | Stochastic Simulation | Artificial Intelligence
	Underlying Technology: Annealing | Superconducting | Trapped Ions

Proteins, composed of amino acids, perform various functions, such as catalyzing reactions and providing body structural support. Protein folding is the physical process in which a protein chain is translated into its native three-dimensional structure with a "folded" conformation to make the protein biologically functional. Researchers aim to design and develop therapeutic molecules that can intervene in the folding process to restore the proper structure and function of misfolded proteins.

There are various ways to fold a protein and link the amino acids along the chain. However, with each additional link on the chain, the problem size increases. Due to the very large number of degrees of freedom in an unfolded polypeptide chain, the molecule has an astronomical number of possible conformations. The check-each-possible-fold approach has a very high computational cost.

Quantum computing can process all viable protein folding patterns to statistically identify the most stable configurations, expediting the drug discovery process by minimizing the need for extensive laboratory experiments.

17.4 USE CASE: DRUG TOXICITY

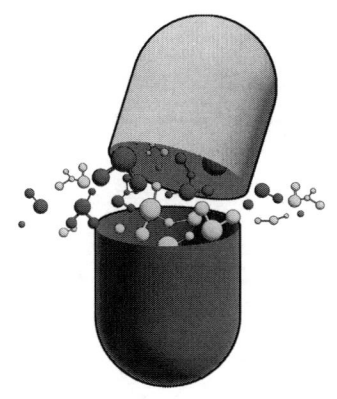

| Business Group: Healthcare & Pharma |
| Problem Category: Objective Optimization |
| Underlying Technology: Annealing \| Trapped Ions |

To avoid clinical trial failure due to adverse drug effects, clearer, more detailed information on potential outcomes and greater explanatory power in interpreting any adverse effects that might emerge, are crucial. Drugs must be analyzed to ensure their safety to the human body by demonstrating they have an acceptable safety profile. Analyzing toxicity predictions for organ systems, appropriate dosing, solubility, and how the human body absorbs a drug is crucial.

The problem is that while clinical trials may tell us that a product is unsafe or ineffective, they rarely tell us why or suggest how to improve it. Moreover, products that fail during clinical trials may simply be abandoned. Predicting low-frequency side effects or the impact of changing the compound bases is difficult because such side effects may not become apparent until the treatment is adopted by many patients. Simulating clinical trials ahead of time is not feasible with current technology.

Quantum computing can examine the structure and chemical characteristics of molecules to foresee potential adverse effects of medications in silico before clinical trials, thereby accelerating the time to market.

CHAPTER 18

Manufacturing

Manufacturing is the transformation of materials or components into new products. Its scope is broad and comprises multiple areas, such as food, textiles, energy solutions, plastics, machinery, and transportation.

It accounts for 16% of the world's GDP, and it has declined over the past decades.

Manufacturing provides 350 million jobs, and workers typically do not need an advanced degree.

Manufacturing's main resources are machine-made or hand-made products such as baked goods, clothing, and confectionaries.

This business group comprises the following types of products:

- **Made-to-stock:** based on year-over-year sales that match inventory with expected demand.

- **Made-to-order:** for customizable products that require customer specification, the production of an item begins only after a confirmed customer order is received.

- **Made-to-assemble:** parts of the product are already manufactured, and they are assembled once the order is received so they can be customized before shipping.

DOI: 10.1201/9781003302674-21

Manufacturing processes involve factors such as training, the work environment, quality control of raw materials, health, and safety. In addition, sustainability issues and related federal regulations are hard to address and add costs of up to $20,000 per employee each year.

Figure 18.1 shows the manufacturing activity flow chart.

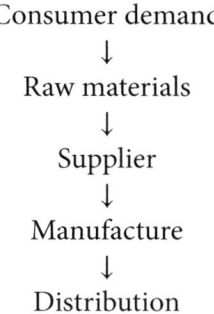

FIGURE 18.1   Manufacturing flow chart.

The manufacturing business group's main pain points are related to the following:

1. Overall, 82% of the equipment efficiency costs industrial plants $10–$250 K per hour.

2. Problems with machinery that slow production increase reactive repair costs and lower productivity.

3. Fluctuating demand for goods creates challenges in enabling cost-effective products. Lean production models reduce waste (defects, overproduction, transportation, non-value-adding processing, motion, waiting, unused talent, and inventory).

4. Quality control, including productivity, overhead, design, and cost-effectiveness contributes to risk mitigation and customer retention.

5. Supply chain costs are increasing, container shipping prices increased between 25% and 50% in recent years with stock breaks requiring bigger provisioning needs.

Some companies testing quantum computing: BASF, Bosch, Boeing, Bridgestone, IAI Automation, Johnson Matthey, OKI, Rolls-Royce, Samsung.

18.1 USE CASE: IMPROVING MATERIALS

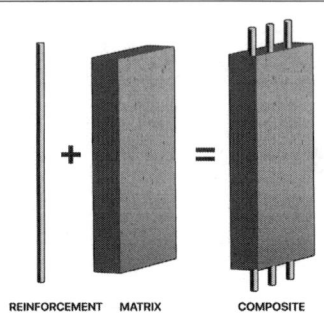

Business Group: Manufacturing | Aerospace | Automotive | Chemicals
Problem Category: Chemical Processes
Underlying Technology: Simulators | Superconducting

Industrial engineers assess materials for their strength, durability, and resilience to determine their suitability for specific applications. Understanding material properties helps ensure product integrity, structural stability, and safety throughout a product's lifecycle. Materials in aircraft, cars, and buildings must be producible in multiple product forms and demonstrate consistent high quality. Apart from low weight requirements, they must have other mechanical properties such as strength, toughness, fatigue life, impact resistance and scratch resistance, especially in aerodynamic situations.

Understanding properties such as weight, strength, thermal resistance, energy efficiency, performance, structural integrity, and durability is key to select the proper materials to create products and avoid product failure. However, material selection, testing, and qualification are long processes. In aerospace, for example, it can take up to 10 years to qualify new materials. The process requires months-long screening in the laboratory, age testing, and accelerated testing of outdoor exposure or in-service evaluation. Companies create climate chambers to test material lifecycles and create cost safety margins.

Quantum computing facilitates the analysis of material properties and reactions, significantly reducing the need for extensive laboratory testing. This improves the competitive edge in material science.

18.2 USE CASE: ASSEMBLY LINE FLOW

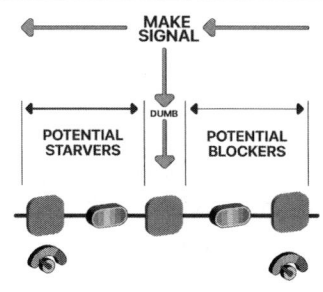

	Business Group: Manufacturing
	Problem Category: Objective Optimization
	Underlying Technology: Annealing

To enhance productivity in a factory, optimizing equipment and material layout and minimizing worker or robotic arm movement distance is crucial to achieve a continuous 'flow' of goods along an assembly line. Productivity improvement involves optimizing schedules, resource allocation, and product mix ratios to reduce costs and save time. This includes improving manufacturing processes, worker efficiency, and resource utilization.

In a factory, the manufacturing process differs for each product, but many pieces of equipment are shared. The worker follows the manufacturing process that describes the order of the processes when manufacturing a product and the equipment used in each process. Products are finished when all manufacturing processes are completed. Optimizing the equipment layout to complete one product can shorten the takt time, i.e., the production rhythm, by up to 30%.

Quantum computing has the capability to optimize the planning of intricate processes with strict time constraints and multiple variables, such as sewing, assembly, screwing, picking, pressing, and die cutting, thereby enhancing efficiency.

18.3 USE CASE: PREDICTIVE MAINTENANCE

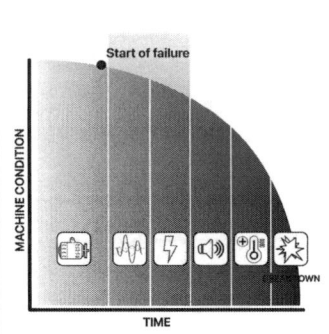

	Business Group: Manufacturing | Aerospace | Automotive | Energy | Telecommunications
	Problem Category: Artificial Intelligence
	Underlying Technology: Superconducting

In manufacturing, monitoring techniques are used to detect failure signals. Real-time monitoring of parameters such as temperature, pressure, vibration, parallelism, and speed ensures they stay within specified ranges. Regular inspections at different production stages involve visual checks, measurements, and functional tests to identify defects or deviations. Statistical process control (SPC) techniques are used to analyze production data for patterns and variations to identify potential failures. Data analysis methods, including statistical analysis and machine learning, are used to find hidden failure signals and correlations, aiding quality control efforts.

Monitoring techniques and data analysis are costly because of the vast amount of data that must be analyzed. An AI dataset can contain more than 800,000 photorealistic images in 80 categories of production resources. Data analysis needs, such as frequency, quality, integration, false positives or negatives, and storage complexity, pose risks, such as direct costs including scrap, rework, warranty claims, or customer returns and indirect costs from production delays, larger maintenance and repair expenses, damaged brand reputation, and potential legal liabilities.

Quantum computing can be employed to predict potential break-downs in machinery, finding hidden data points, thereby increasing the reliability of production processes and preventing unexpected equipment failures.

18.4 USE CASE: COMPONENTS PERFORMANCE

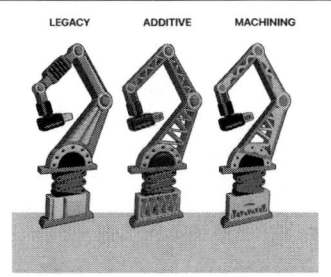

LEGACY ADDITIVE MACHINING	Business Group: Manufacturing | Aerospace | Automotive
	Problem Category: Stochastic Simulations
	Underlying Technology: Annealing | Superconducting

Light weighting design to develop products and structures that maximize strength and functionality while minimizing weight is extensively used in manufacturing. This includes considering factors such as load-bearing capacity, performance, safety regulations, and environmental considerations. It involves analyzing materials, generating design concepts, conducting structural analysis and simulations, and optimizing feasibility and cost in manufacturing.

Lightweight design must be compatible with the chosen manufacturing processes, balancing various factors simultaneously, such as material structural integrity, performance requirements, manufacturability, material selection, and cost. Components based on new materials must be analyzed in a holistic manner, considering all relationships within the vehicle and their manufacturing. The impact of producing nonperformance components can have lethal costs, such as fatalities and the recall of millions of products. For an aircraft such as the Boeing 787, a 20% weight savings results in 10% improvement in fuel efficiency.

Quantum computers can expand the range of design possibilities prior to production running many more simulations. This aids in optimizing material properties, increasing productivity and sustainability.

CHAPTER 19

Supply Chain

Supply chains combine logistics with transportation to manage and deliver goods from manufacturing to end customers, forecast demand, update it, and manage information flow along the entire chain.

The supply chain management market is worth approximately $28.9 billion with an expected compound annual growth rate of 0.4%, generating 450 million jobs worldwide.

Supply chains resources cover producers, vendors, warehouses, transportation firms, distribution centers, and retailers.

This business group comprises activities to ensure resources are delivered on time.

- Demand forecasting involves predicting future demand for a product.

- Warehouses provide goods storage and materials and customize the final product package.

- Inventory management is used to maintain the optimum number of items, pick and pack items, and prepare goods prior to being shipped.

- Shipping is performed in different methods to deliver the package to the client.

- Managing customer returns includes reverse logistics and asset recovery.

DOI: 10.1201/9781003302674-22

The global logistics and supply chain industry market is moderately fragmented, with several companies of various sizes and major companies accounting for the market in specific parts, which impacts economic trading since many containers return to the export country empty.

Figure 19.1 shows the supply chain activity flow chart.

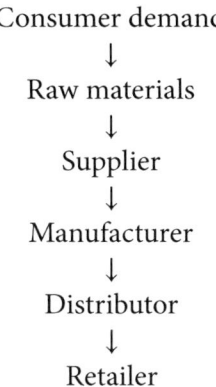

FIGURE 19.1   Supply chain flow chart.

The supply chain business group's main pain points are related to the following:

1. Quality control across the supply chain, including suppliers, must be connected to share information precisely. Disruptions can cost $440–$660 million in lost sales per week.

2. Shipment tracking services need to be available to track components/shipments en route, with the ability to notify people in case of disruptions and provide estimated arrival times.

3. Inaccuracy in predicting and adapting to demand changes impacts fulfillment rates and causes missed revenue targets and stock-outs.

4. The packaging process and greenhouse gases can cause environmental damage through transportation, and container optimization is lacking, which requires considerable energy.

5. Numerous variables are managed by suppliers and geographic locations include different countries to narrow the gap between consumers and production.

Some companies testing quantum computing: Bank of England, BMW, Delta Airlines, DHL, Eneos, FedEx.

19.1 USE CASE: ENERGY DELIVERY

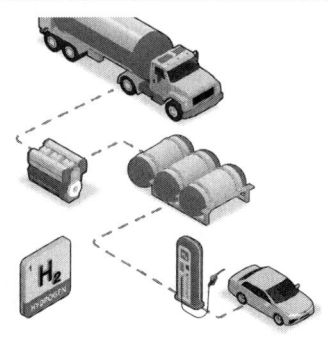

	Problem Group: Chemical Processes
	Business Category: Supply Chain | Energy | Government
	Underlying Technology: Trapped Ions | Superconducting

To achieve the objectives of the EU Hydrogen Strategy, countries must have access to renewable hydrogen, which may come from different locations where renewable electricity is less expensive. Hydrogen must be transported from the point of generation to the point of use. The supply chain process covers its production with different methods; its compression packaging to increase its density by reducing its volume and making it easier and cheaper to transport; and storage in containers, such as high-pressure tanks or pipelines for distribution to end users.

Hydrogen distribution is complex since it is a substance that permeates and is highly flammable. This, together with its low volumetric energy density, makes it difficult to store and transport. It is crucial to reduce its transportation cost, increase its energy efficiency, maintain its purity, and minimize leakages. To transport it, hydrogen must be pressurized and delivered as a compressed gas or liquefied.

Quantum computing can improve different clean energy transportation methods by analyzing chemical reactions for better catalysts to pressurize hydrogen and better transport it as compressed gas or liquid, improving security, costs, and sustainability.

19.2 USE CASE: LOAD OPTIMIZATION

	Business Group: Finance | Logistics | Supply Chain
	Problem Category: Artificial Intelligence
	Underlying Technology: Annealing

It is crucial to keep just enough cash in ATMs and banks to provide the service customers expect while moving cash as little as possible in the network and keeping as much as possible in bonded stores. Cash replenishment accounts for between 35% and 60% of the total cost of operating an ATM. It is a dangerous process to transport money.

Calculating the ideal amount of cash to be held in any individual ATM at one time and determining the most effective route for machine-by-machine cash replenishment is a complicated exercise. The possible permutations of amounts and routes across an ATM network are daunting, and calculating the optimum solution even for a single-bank-owned network within a timeframe short enough to be able to respond on the ground has been found to be unrealistic.

Quantum computing can analyze data to perform projections on cash distributions more precisely to optimize replenishment schedules and routes and minimize the costs of cash transport, storage, and security.

19.3 USE CASE: JUST-IN-TIME LOGISTICS

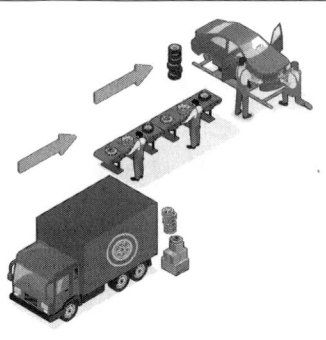

Business Group: Automotive | Supply Chain	
Problem Category: Objective Optimization	
Underlying Technology: Trapped Ions	

On average, vehicles consist of 30,000 components, and many of these are built from various components that must be moved from material to vehicle installation and then on to the customer. Logistics ensures that materials, goods, and services flow smoothly between wholesalers, distributors, retailers, and customers, meeting their demand, coordinating all involved parties, and ensuring providers arrive just in time.

Raw materials are shipped to a refinery which then ships them to a component supplier, who builds the component using other materials. The part is then shipped to an engine supplier who puts it in a new unit and then ships it to a car factory where it is put into a vehicle and sent to a dealership to be given to a customer. Between each of these steps, however, there are logistics and storage to consider. Blind spots at any stage of manufacturing can impact the entire procurement process and result in the loss of billions of dollars every year.

Quantum computing algorithms can reduce the size of the combinatorial problem, improving the flow of materials and goods, reducing bottlenecks, and improving production, delivery, and customer satisfaction.

19.4 USE CASE: DEMAND FORECAST

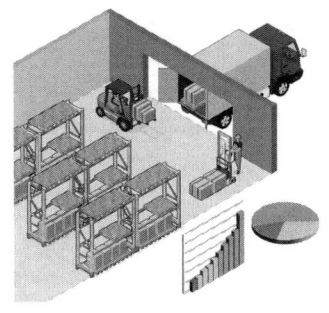

	Business Group: Supply Chain	Energy
	Problem Category: Stochastic Simulation
	Underlying Technology: Annealing | Superconducting

In supply chains, future product demand predictions based on historical data, market trends, and other relevant factors are crucial for production planning, inventory management, and future capacity requirements, among other things. They help determine the expected demand levels, seasonal patterns, and potential fluctuations, providing a roadmap to anticipate and adapt to changing conditions and minimizing risks.

Demand forecasting is performed by incorporating uncertainty and variability to generate multiple scenarios by randomly sampling from probability distributions to identify potential outcomes and most likely risks. Forecasting inaccuracies affect the entire supply chain, creating a bullwhip effect that causes progressively larger fluctuations, resulting in wasteful and costly surplus, failure to meet customer needs, loss of opportunities, and eventually, $100 million sales losses.

Quantum computing algorithms are adept at managing the probabilistic aspects of demand. Improved forecasting through these algorithms can optimize sales and revenue, minimize excess inventory, and free up storage space.

CHAPTER 20

Telecommunications

Telecommunications is defined as communicating over a distance. It plays a critical role in connecting people, businesses, and economies.

Every 10% increase in a country's connectivity score leads to a 1.28% increase in its GDP.

It is very capital intensive, with high R&D costs and continuous capital reinvestment.

Telecommunications' main resources are mobile voice, messaging and data, leased lines, private networks, fixed and broadband, unified communications, and web conferencing.

This business group comprises activities to ensure proper connectivity service quality. It comprises the following activities:

- Cable companies that provide television, internet access, and phone services using underground cables.

- Internet service providers that give access to the internet and its surrounding solutions.

- Satellite companies that create a communication channel between a source transmitter and a receiver at different locations on Earth.

- Telephone companies provide voice and data telecommunication services.

DOI: 10.1201/9781003302674-23

Telecommunications has evolved from just a few large players in the market to a decentralized market which is volatile and competitive, to win subscribers and customers.

Figure 20.1 shows the telecommunications activity flow chart.

Consumer device
↓ ↑
Transmission Lines
↓ ↑
Switching facilities
↓ ↑
Network operation
↓ ↑
Network backbone

FIGURE 20.1   Telecommunications flow chart.

The telecommunications business group's main points are related to the following:

1. With the number of devices connected growing, network encryption is a challenge.

2. Network management with antenna deployment planning is a complex problem; in some other areas, overlap between signals cause a waste of resources.

3. Low transmission quality results in poor streaming.

4. Implementing 5G would allow efficient management of the devices connected to the network. Additionally, since the network needs constant monitoring to operate properly, this solution would allow to configure the 5G network in almost real time.

5. GPS localization in occasions does not work properly on certain areas or is not precise enough for the task needed.

Some companies testing quantum computing: BT, Cellnex, Ericsson, LG, Nokia, Telecom Italia, Orange, South Korea, Vodafone.

20.1 USE CASE: RADIO ACCESS

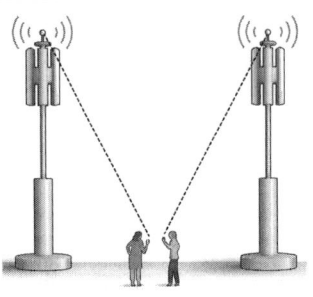

Business Group: Telecommunications
Problem Category: Chemical Processes
Underlying Technology: Superconducting

A Radio Access Network (RAN) is a mobile telecommunication system component implementing a radio access technology to connect mobile devices or computers with the core network. It sends information via radio waves from end-user devices to a RAN's transceivers, and from there to the core network, which connects to the global internet. RAN technology has advanced tremendously since it was first introduced, and it is still used today in modern 5G networking.

RANs are crucial connection points that represent significant overall network expenses, perform intensive and complex processing, and generate heat inside. As the demand for data increases, the physical layer processing functions for transmission needs cooling to work efficiently. Carriers need to be switched off to save energy based on the client usage seasonality analysis, which can bring ~15% in energy savings.

Quantum computing algorithms have the capability to expedite the resolution of linear equations, thereby enabling RANs to manage data more effectively. This can lead to increased revenues and a reduced environmental footprint.

20.2 USE CASE: NETWORK PLANNING

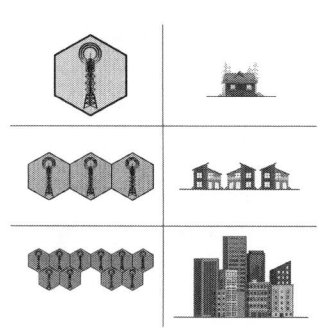

	Business Group: Telecommunications
	Problem Category: Objective Optimization
	Underlying Technology: Annealing

Cellular networks are distributed over land areas called cells. Each one is served by a transceiver, which are joined together to provide radio coverage over a specific geographic area. Each cell is assigned to multiple frequencies with corresponding radio base stations that can be reused by other cells if they are available. Network design and topology are essential to provide reliable coverage and capacity.

Since the same frequencies can be used by multiple cells, a common problem in network design is crosstalk from two different radio transmitters using the same channel or an overlapping channel. This leads to signal degradation and reduces the effective bandwidth within the affected channels, limiting the overall network capacity. This limitation can require additional network infrastructure that is not aligned with client volume.

Quantum computing can select the best frequencies to use, augmenting network capacity and reliability. This can result in cost savings and provide a competitive advantage as demand for network services grows.

20.3 USE CASE: SERVICE QUALITY

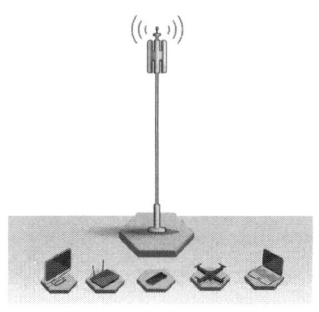

	Business Group: Telecommunications
	Problem Category: Artificial Intelligence
	Underlying Technology: Photonics | Superconducting

Communication networks need to monitor user behavior to detect anomalies, enhance performance, adapt to changing conditions, and make decisions. To do this, the location of mobile devices is crucial, as is interpreting information from network sensors to identify potential issues such as congestion before they escalate and to enable rerouting of traffic to prevent outages. This approach optimizes resource allocation, boosts network efficiency, and improves service quality.

As wireless networks grow in complexity, many factors affect the quality of wireless communications, and traditional computing resources cannot cope with the computational demands to properly predict the quality of user experience for video streaming based on device and network-level metrics. Current methods use statistical techniques with poor accuracy (approximately 70%) and lack the ability to evolve prediction models with the system's dynamics, which negatively impacts user experience.

Quantum computing can enhance network performance by creating adaptable models that respond to environmental changes. By analyzing multiple data points, models can prevent outages and reduce energy consumption.

20.4 USE CASE: MIMO SPECTRUM EFFICIENCY

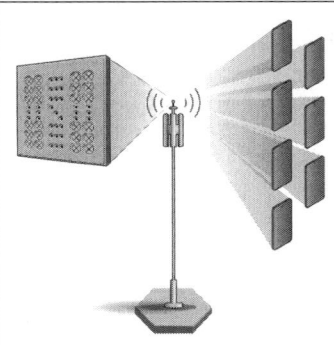

	Business Group: Telecommunications | Government
	Problem Category: Stochastic Simulations
	Underlying Technology: Annealing

As the future network infrastructure is deployed, there is an ever-increasing data traffic and subscriber growth with new performance demands. Networks will require new levels of spectral efficiency and flexibility. Massive Multiple-Input, Multiple-Output (MIMO) technology uses multiple antennas to send and receive data to improve channel capacity, allowing for higher data rates. It equips base stations with a very large number of antenna elements, on the order of tens, hundreds, or even thousands of antennas in a single antenna array to improve spectral and energy efficiency.

Designing massive MIMO technology is a complex task that requires many antenna elements, which increases the cost and power requirements of the system. It also requires proper calibration to achieve channel reciprocity. Modeling, simulating, and testing MIMO is difficult since deployed physical prototypes do not exist and must be simulated. Proper MIMO simulation can increase network capacity and cell edge use throughput by 40%.

Quantum computing can refine the MIMO algorithm, reducing its computational complexity with multiple simulations. This boosts network capacity and quality of service.

CHAPTER 21

Use Case Problem Mapping

O NCE WE HAVE CHOSEN a use case to test with quantum computing, we need to map the underlying problem onto a quantum algorithm. The first thing that must be done is to understand the problem well so that it can be compared to a similar classical problem that can be used to obtain a good mathematical intuition of the problem and provide a benchmark to delimit its values.

21.1 EXAMPLE PROBLEM: PACKAGE DELIVERY

This problem tries to answer how to find the optimal route for a vehicle to deliver packages to a set of customers. The problem can have different modalities:

1. Multiple-route customer-visit on which the driver returns to the warehouse after each route to pick up more packages for the next route. This is called the vehicle routing problem (VRP).

2. One-route customer-visit to deliver packages starting and ending from the warehouse since there is only one loading before visiting all customers, which is especially important for those cases when returning to the warehouse is not feasible. This is called the traveling salesman problem (TSP).

Figures 21.1 and 21.2 illustrates both problems, showing the difference between them.

 DOI: 10.1201/9781003302674-24

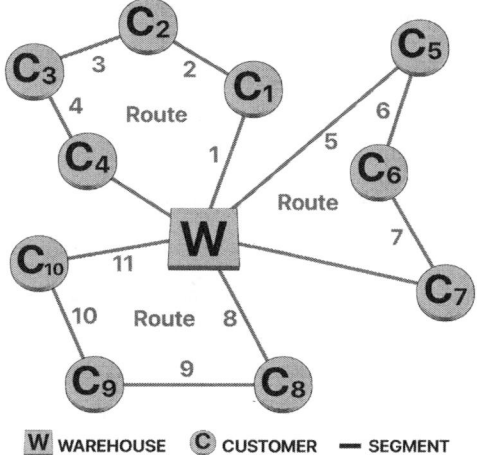

FIGURE 21.1   VRP.

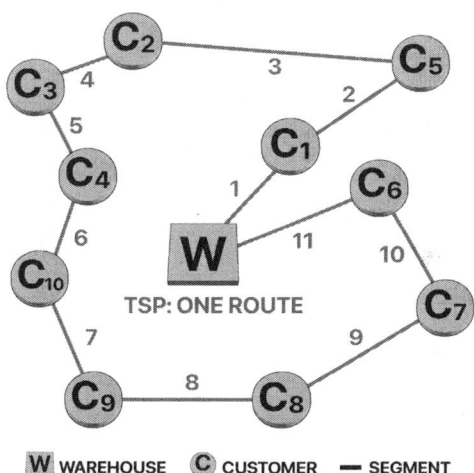

FIGURE 21.2   TSP.

The problem is represented mathematically by a graph with:

- A node that represents a customer to be visited.

- A warehouse that contains all the packages to deliver.

- The segment distances between different places.

Finding this optimal route that allows to visit all customers in one go is a complex graph mathematical problem, since there are different potential route combinations for possible routes to deliver the packages, in particular, there are $c!$ combination routes, whereas c grows, the number gets huge and highly complex to calculate.

By reducing the VRP to a TSP, a well-known problem, we create a cycle graph problem with less possible cities. If the graph is directed with a specific travel direction, then the number of possible routes is $(c-1)!$ otherwise it is halved $\dfrac{(c-1)!}{2}$, as it is undirected, each route and its reverse are considered the same.

Using a brute-force approach, we also list all the possible routes to analyze the distance between them by calculating the segment size from all customer combinations. This helps to establish lower and higher bounds that can be a benchmark for routing length optimization. We visually illustrate this approach on Table 21.1.

TABLE 21.1	Illustrative Locations Distance Heat Map				
	C1	**C2**	**..**	**C8**	**C9**
C1		best			
C2	best				
:					best
C8			best		
C9				best	
0	+	++	+++	++++	+++++
Note: Distance between locations from short to long					

Mapping the problem onto a quantum algorithm can be approached in two different ways depending on how the variables are encoded:

1. Qubit encoding that represents each binary variable encoding whether or not a given customer is visited at a given time step, into the qubits which is $O\ (c^2)$.

2. Amplitude encoding uses the 2^n amplitudes of n qubits to encode the $c!$ possible paths through the network, which may result in $\log_2 (c!)$ qubits (since it is $2^n = c!$)

Amplitude encoding allows to leverage amplitude amplification with faster results than a classical brute force search would do, obtaining a quadratic speedup that reduces the polynomial growth of the problem to a non-factorial number.

When selecting use cases, it is important to assess qubit requirements, which affect quantity and performance that influence information storage and processing capabilities. Currently, there is no standard method to determine exact qubit needs, however there is an intuition based on the type of data processed as seen in Table 21.2.

TABLE 21.2 High-Level Qubit Needs Estimate for Programming		
Variables	→	**Qubits**
When calculating the electronic structure of molecular systems, which are used to analyze chemical properties and interactions. Variables (i.e., molecule spin orbitals) are mapped into qubits. After that, the gates are used to analyze the energy levels towards a specific molecular value. The larger the molecule, the longer the number of operations, hence gates, needed.	1:1	The variables are mapped individually since they use both 1 and 0 values of the qubits.
When calculating the optimal value of graphs that represent relations between nodes, which are used for problem solving. Variables (i.e., graph nodes) are mapped into qubits. After that the gates are used to analyze the paths. The bigger the problem is with many interrelations and nodes, the greater the number of operations, hence gates, needed.		
When calculating linear equations to analyze relationships between two variables that are used for forecasting or AI. Variables (i.e., number of unknowns) are logarithmically mapped into qubits. After that, the gates are used to perform all the matrix calculations. The greater the number of unknown values, the longer the number of operations, hence gates, needed.	Log N	The variables are mapped into the different qubit states, using the 2^n advantage.
When analyzing stochastic models to perform random sampling to find numerical estimates on mathematical and physical problems. Variables (i.e., number of samples) are logarithmically mapped into qubits. After that, the gates are used to perform the repeated samples in the probability distribution. The larger the number of variables, the more sampling, and operations, hence gates, needed.		

CHAPTER 22

How to Get Quantum Ready

R ATHER THAN WAITING FOR quantum computing to fully mature, enterprises should concentrate on testing the technology and becoming ready as soon as possible. We have already reviewed how they should get ready for the downsides of quantum computing security risks by creating a strategy to update their encryption schemes.

In this book, we have seen some of the foundational concepts of quantum computing, and you may have already realized that the learning curve is a complicated one, and there is a lot of physics involved in the programming. We are not migrating from one technology to another, like we have done in the past. We are rethinking the problems from a quantum lens, applying the concepts of quantum mechanics we have reviewed in this book and using quantum algorithms that are being evolved as we speak.

Approaching quantum computing is not an easy task. The learning curve is quite steep and deciding to "get quantum ready" requires making an informed decision for which we need to analyze the external ecosystem, our internal capacities, and the technology to use. This preliminary evaluation will help companies define the road to follow and build a business case before engaging in any expensive implementation plan.

To create a quantum computing roadmap, companies need to evaluate their current and future situation, to understand the specific needs in terms of people, estimated budget, potential providers, use cases that can

DOI: 10.1201/9781003302674-25

benefit from the technology and timeframes. This can be done by answering the questions detailed in Table 22.1.

TABLE 22.1		Quantum Readiness Framework
1	**Strategy:**	Where am I now and how far (involved) do I want to go?
2	**Team:**	Can people be dedicated to this, and do I have the necessary skills?
3	**Uses:**	How to define, analyze, prioritize, and select use cases?
4	**Provider:**	What type of provider is the best for me and how to select it?
5	**Tests:**	Which technology will I need to build and evaluate the use cases?

22.1 STRATEGY: WHERE AM I NOW AND HOW FAR DO I WANT TO GO?

Companies need to get a grasp of the quantum computing ecosystem (shown on Figure 22.1), understanding who the main players are, what are the software tools, what hardware works best, what are the use cases being tested in each industry, and what is the funding situation from the government and Venture Capitalists. It helps to attend a quantum computing industry and a scientific international conference to get familiar with different players.

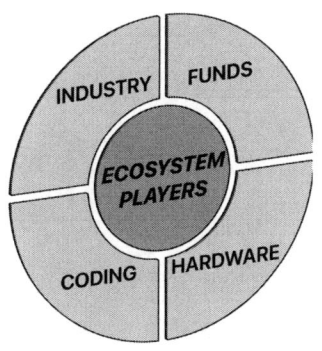

FIGURE 22.1   Ecosystem players.

- Coding: Development tools accessible online and compatible with existing programming languages.

- Funds: Public or private investment from government or Venture Capitalists.

- Hardware: Fast access to existing computers online to integrate with existing infrastructure.

- Industry: Applicability use cases that have potential business advantage.

22.2 TEAM: DO PEOPLE HAVE THE TIME AND SKILLS TO WORK ON THIS?

When deciding to start implementing a quantum computing adoption plan, we need to identify if there are people who can work on the project, which will require full-time dedication, do they have the necessary knowledge? If not, can they be upskilled through specialized courses? To answer this, it is useful to know what skills are needed to work on quantum computing, which are shown on Figure 22.2.

- Software coding: Developers that are versed in Python or other programming language.

- Linear algebra: Learned in STEM university degrees, a certain advanced level is necessary.

- Quantum physics: Part of some physics and engineering university curriculum.

- Quantum information: very specific to physics degrees education.

FIGURE 22.2   Necessary skills.

22.3 USES: HOW TO DEFINE, ANALYZE, PRIORITIZE, AND SELECT USE CASES?

It is first necessary to identify existing inefficiencies in companies' operations-specific use cases and evaluate where quantum computing could provide support, thus companies need to:

1. Analyze which is the enabling algorithm and its problem category.

2. Estimate the expected business impact and monetary return.

3. Identify the computation limitation or inefficiency source.

Enterprises need to understand that for many reasons, such as not having the necessary technology to solve a problem properly, or analyzing large amounts of complex data that they cannot process, or not having enough time to run an algorithm, they make informed approximations that are not optimal.

Once these use cases are analyzed, the next step is to evaluate the underlying algorithm's expected advantage, and its performance according to the technology evolution. In order to do this, it is necessary to understand the technological requirements of the algorithm in terms of its resistance to noise.

Finally, a key factor is the "size of the prize" that is, the potential gain from improving the use case through quantum computing.

With all these analyses we can map a use case portfolio onto a prioritization matrix that will help to make an informed decision on a quantum adoption strategy.

Table 22.2 shows the prioritization matrix framework, the bubble size illustrates the business potential, and the Y axis and X axis analyze the advantage and performance.

22.4 PROVIDER: HOW TO SELECT THE BEST PROVIDER FOR ME?

When selecting a technology to work with, the process is slightly different from selecting a regular technology, since there are nuances in understanding how hardware works and how to map an industry use case onto its mathematical interpretation.

Providers must have in-depth quantum computing knowledge, from software development to hardware requirements, with specifics on how to handle qubits and must have business analysts to understand the industry extremely well and have the capacity to contextualize the problem to a similar mathematical one.

Another question to consider is if we should work with a provider that has the following characteristics:

- Offers an integrated approach of hardware with software.

- Can develop over different hardware platforms.

- Brings expertise on specific industry use cases.

It is advisable that companies start with small-scale experiments and development projects to gain hands-on experience with quantum computing.

TABLE 22.2   Use Case Prioritization Framework

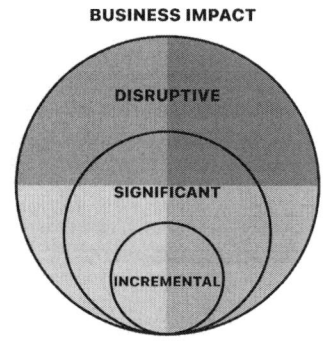	Analyze the algorithm used to solve the use case to understand its quantum advantage and technical performance. • **Quantum advantage:** theoretical improvement the algorithm brings in relation to a classical algorithm. • **Technical performance:** how well does the algorithm work with the existing technology and what type of error correction does it need to work well.
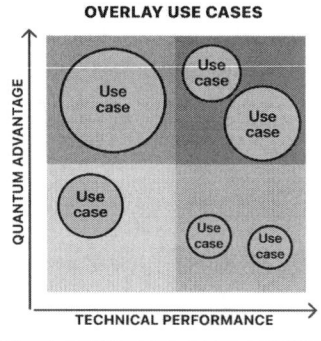	Evaluate use case business impact, which can for example be split into three values, that can be described by the economic value they bring, as defined earlier: 1. incremental ~US\$B [0–30] 2. significant ~ US\$B [30–60] 3. disruptive ~ US\$B [60–90]
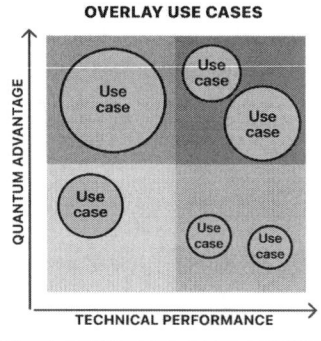	Map each use case with its underlying algorithm and place it on the corresponding matrix quadrant, with the business value as an independent layer on top of the matrix.

22.5 TESTS: WHICH TECHNOLOGY WILL I NEED TO SELECT TO EVALUATE THE USE CASES?

Each problem has different programming requirements that hardware must address which includes an architecture that affects data loading, gate speed, and result readout. There is no clear hardware performance benchmark, but there are some criteria to consider which we have seen on the qubit analysis section and explain them in Figure 22.3.

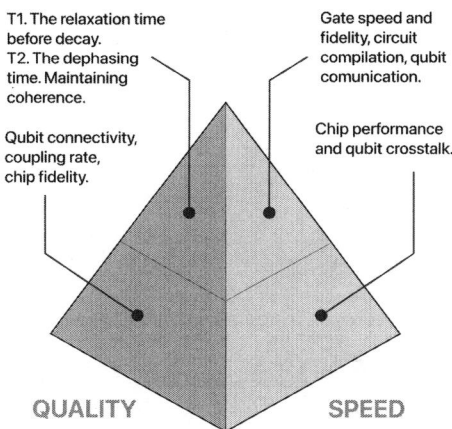

- Maintaining qubit **quality** throughout the whole operation and as the number of needed qubits grows.

- Achieving high computational **speeds** to allow to perform all the necessary operations regardless of the computation complexity.

FIGURE 22.3   Qubit technology evaluation.

With these considerations, it is important to choose the type of quantum computing model to use (gate-based quantum computing or quantum annealing), how easy it will be to access it via the cloud and to integrate it with the corporation's technology infrastructure.

CHAPTER 23

Responsible Technology Use

S CIENCE AND TECHNOLOGY PLAY key roles in enhancing productivity and advancing society by bringing advancements, that improve the quality of life. However, there are potential misuses that can exacerbate current ethical dilemmas and security risks in the rapidly evolving technological landscape.

There are three key categories of good practices to ensure that advancements in science and technology contribute to a more inclusive, sustainable, and peaceful world. These practices are explained in Table 23.1.

TABLE 23.1	Quantum Computing Ethical Good Practices
Inclusive	Research is primarily conducted by leading academic institutions in certain nations, concentrating on high-tech knowledge and creating a geographical divide even within advanced economies. Limited access to knowledge and funding reduces inclusivity in the field. Additionally, varying levels of digital maturity among countries limit their capacity to work on advanced technologies. This disparity also applies to the different technology adoption speeds between innovative and laggard companies.

(Continued)

DOI: 10.1201/9781003302674-26

TABLE 23.1 (*Continued*) Quantum Computing Ethical Good Practices	
Sustainable	Quantum computers are expected to use significantly less energy than supercomputers, as in Section 3.2 of this book, because of their operational principle, this implies as well that their energy requirements do not grow at the same rate as their computing capacity. However, they still require substantial energy to be built and operate effectively today. Additionally, constructing a quantum computer may exacerbate the current scarcity of critical raw materials, such as helium, lithium, niobium, nickel, platinum, and gallium. On the other hand, quantum computing holds the potential for creating more ecological materials and energy sources that can contribute to improve sustainability.
Peaceful	The nature of technology makes it ideally suited for the creation of both cyber and physical weapons, potentially enabling misuse by those seeking to harm or dominate. Examples include security codebreaking using algorithms like Shor's factorization, the misuse of chemical processes to create harmful molecules such as toxins, and the acceleration of intelligent machine development for malicious purposes.

Disruptive technologies have significant implications, and policymakers must be prepared to respond. By learning from past high-tech experiences to apply effective strategies to quantum technology. Here are some key aspects to consider for developing and running quantum technology solutions to avoid ethical pitfalls:

23.1 DISTRIBUTE BENEFITS EQUITABLY

Ensure the benefits of quantum technology are shared globally to prevent widening existing socio-economic gaps.

23.2 REDUCE ENVIRONMENTAL IMPACT

Minimize the environmental concerns from the substantial energy required for quantum hardware, considering the following aspects:

- Public funding for research to lower energy requirements.

- Carbon taxes to encourage less energy-intensive solutions.

23.3 ENSURE ALGORITHM ACCOUNTABILITY

Prevent the 'Black Box Effect' by ensuring transparency in how algorithms function and make decisions. This helps to understand how decisions are made by stakeholders, so that there can be accountability if there are output errors.

23.4 RESKILL EMPLOYEES

As quantum computing advances and replaces existing technologies, reskill employees across all levels, including developers, managers, and executives, to mitigate job impacts.

23.5 PREVENT HOSTILE USE

Implement measures to prevent the use of quantum computers to break existing encryption protocols, which could happen within a decade. In the meantime, ensure that all countries and corporations are updating their encryption techniques to make them quantum safe, as explained in Section 8.3 of this book.

23.6 SECURITY DILEMMA

Prevent competition among nations in developing advanced weapons and surveillance systems using quantum technology to avoid a military arms race. It is crucial to understand the military applications of quantum technology and how it may be destabilizing, such as decrypting messages, enhancing battlefield intelligence through intelligent satellites for new monitoring capabilities, or turbocharging AI military applications. The use of quantum technologies should be included in arms control agreements.

Quantum computing can boost the Sustainable Development Goals (SDGs). It can improve some science and technology evolution, such as create better chemical compounds, simulate different scenarios with stochastic data, optimize management processes, or enable a deeper understanding of complex problems leveraging AI. It also presents a chance to improve society-related topics as the ecosystem is being built.

- **Goal 1:** End poverty – Increase productivity through various solutions that quantum computing can bring.

- **Goal 2:** Zero Hunger – Promote more sustainable agriculture by simulating organic nitrogen fixation processes.

- **Goal 3:** Good Health and Well-Being – Enhance drug development processes through improved protein folding analysis, making medicines more accessible.

- **Goal 4:** Quality Education – Improve access to education programs by updating curricula and reskilling, facilitated by quantum technologies.

- **Goal 5:** Equality of Opportunity by Gender – Ensure that education and workforce opportunities include women and girls to achieve gender equality and empower all.

- **Goal 6:** Clean Water and Sanitation – Simulate various path trajectories to make clean water available in different areas.

- **Goal 7:** Affordable and Clean Energy – Foster access to green power resources with new chemical catalysts developed through quantum computing.

- **Goal 8:** Decent Work and Economic Growth – Include all countries in the development of quantum technology to foster global economic growth.

- **Goal 9:** Industry, Innovation, and Infrastructure – Improve industrialization by providing access to quantum computing infrastructure.

- **Goal 10:** Reduced Inequalities – Ensure cloud access to quantum technologies and solutions built so that all countries can benefit equitably.

- **Goal 11:** Sustainable Cities and Communities – Optimize urban management specially people and product transportation.

- **Goal 12:** Responsible Consumption and Production – Create sustainable processes and improve offer-and-demand forecasts for just-in-time production.

- **Goal 13:** Climate Action – Optimize production processes and create more sustainable materials to combat climate change.

- **Goal 14:** Life Below Water – Develop biodegradable materials and energy sources that do not harm aquatic ecosystems.

- **Goal 15:** Life on Land – Optimize resource management and simulate different degradation scenarios to act beforehand.

- **Goal 16:** Peace, Justice, and Strong Institutions – Ensure quantum technologies are used for ethical purposes and to promote peace and justice.

- **Goal 17:** Partnerships for the Goals – Foster international collaborations between private and public institutions to achieve these goals.

Resources

This section provides some resources that can be useful for subject matter experts, researchers, developers, and students, both from the scientific and industry world.

BOOKS

A Short Introduction to Quantum Information and Quantum Computation by Michel Le Bellac. Cambridge University Press. 2006

An Overview of Quantum Computing: The State of the Art in Computers " by Paul F. Kisak (Self Edit). 2016.

Elements of Quantum Computing. History, Theories and Engineering Applications by Seiki Akama. Springer. 2015.

Explorations in Quantum Computing by Colin P. Williams. Springer Telos, 1st ed. 1998.

Quantum Algorithms via Linear Algebra: A Primer by Richard J. Lipton and Kenneth W. Regan. The MIT Press. 2014.

Quantum Computation and Quantum Information by M. A. Nielsen and I. L. Chuang. Cambridge University Press, 1st ed. 2000.

RESEARCH PAPERS

"Information and Computation: Classical and Quantum Aspects" by Alberto Galindo and Miguel-Angel Martin-Delgado. Rev. Mod. Phys. 2000.

"Quantum Information Theory" by Charles H. Bennett and Peter W. Sho . IEEE. 1998.

"Secure Quantum Key Distribution with Realistic Devices" by Feihu Xu et al. APS. 2020.

"Simulating Physics with Computers" by Reichard P. Feynman. Springer Link. 1982.

REPORTS & WHITE PAPERS

"The Next Decade in Quantum Computing—and How to Play" by BCG. 2018 https://www.bcg.com/publications/2018/next-decade-quantum-computing-how-play

"Quantum Manifesto for Quantum Technologies" by Marcus, Freeke Heijman, Ignacio Cirac, Richard Murray, and Tommaso Calarco. European Commission. Futurium. 2016.

"Quantum Computing Is Becoming Business Ready" by BCG. 2023. https://www.bcg.com/publications/2018/next-decade-quantum-computing-how-play

"Quantum Technology Monitor" by McKinsey. 2023. https://www.mckinsey.com/~/media/mckinsey/business%20functions/mckinsey%20digital/our%20insights/quantum%20technology%20sees%20record%20investments%20progress%20on%20talent%20gap/quantum-technology-monitor-april-2023.pdf

"Technology Quarterly. Here, There and Everywhere" by The Economist. https://www.economist.com/news/essays/21717782-quantum-technology-beginning-come-its-own

WORLD STATISTICS

https://ourworldindata.org, https://data.worldbank.org, https://www.statista.com, https://stats.oecd.org

SDKS

Azure Quantum by Microsoft: https://azure.microsoft.com/es-es/products/quantum
Braket by Amazon: https://aws.amazon.com/braket/
Cirq by Google: https://quantumai.google/cirq
CUDA Quantum by NVIDIA: https://nvidia.github.io/cuda-quantum/latest/index.html
Forest by Rigetti: https://qcs.rigetti.com/sdk-downloads
LabOne Q by Zurich Instruments: https://www.zhinst.com/europe/en/quantum-computing-systems/labone-q
Ocean by D-Wave: https://docs.ocean.dwavesys.com/
PennyLane by Xanadu: https://pennylane.ai
Qiskit by IBM: https://www.ibm.com/quantum/qiskit
Tket by Quantinuum: https://tket.quantinuum.com

SIMULATORS

Intel QS by Intel: https://intel-qs.readthedocs.io/en/docs/index.html
QuASK by CERN: https://quask.web.cern.ch
Quirk by Craig Gidney: https://algassert.com/quirk

Bibliography

"2020 Airbus Quantum Computing Challenge" Airbus web. 2020 [WEB POST]. https://www.airbus.com/en/innovation/disruptive-concepts/quantum-technologies/2020-airbus-quantum-computing-challenge

"A BBVA and Zapata Computing Study Shows the Potential of Quantum Computing for Derivative Calculations" BBVA Website. 2021 [WEB POST]. https://www.bbva.com/en/innovation/a-bbva-and-zapata-computing-study-shows-the-potential-of-quantum-computing-for-derivative-calculations/

"A Biological Sequence Comparison Algorithm Using Quantum Computers" *Nature*. 2023 [RESEARCH PAPER]. https://arxiv.org/pdf/2303.13608.pdf

"A Comprehensive Review of Quantum Random Number Generators: Concepts, Classification and the Origin of Randomness" by Vaisakh Mannalath et al. Springer Link. 2023 [RESEARCH PAPER]. https://link.springer.com/article/10.1007/s11128-023-04175-y

"A Conceptual Architecture for a Quantum-HPC Middleware" by IEEE. 2023 [RESEARCH PAPER]. https://arxiv.org/pdf/2308.06608.pdf

"A Fast Quantum Mechanical Algorithm for Database Search" by Lov K. Grover. ACM. 1996 [RESEARCH PAPER]. https://dl.acm.org/doi/pdf/10.1145/237814.237866

"A Fast Quantum Mechanical Algorithm for Database Search" by Lov K. Grover. ACM. 1996 [RESEARCH PAPER]. https://dl.acm.org/doi/pdf/10.1145/237814.237866

"A Fertilizer Revolution is on the Horizon" by Alberta Farmer Express. 2021 [WEB POST]. https://www.albertafarmexpress.ca/news/a-fertilizer-revolution-is-on-the-horizon/

"A Perspective on Protein Structure Prediction using Quantum Computers" by Hakan Doga et al. *Inspire High-Energy Physics*. 2023 [RESEARCH PAPER]. https://arxiv.org/pdf/2312.00875.pdf

"A Quantum Computing Approach for the Unit Commitment Problem" Springer. 2022 [RESEARCH PAPER]. https://arxiv.org/pdf/2212.06480.pdf

"A Quantum-Inspired Approach to De-Novo Drug Design" Fujitsu. [WHITEPAPER]. https://www.fujitsu.com/fi/imagesgig5/Healthcare-Assets-Whitepaper.pdf

"A Quantum–Quantum Metropolis Algorithm" by Man-Hong Yung and Alán Aspuru-Guzik. *PNAS*. 2012 [RESEARCH PAPER]. https://www.pnas.org/doi/10.1073/pnas.1111758109

"A Quantum–Quantum Metropolis Algorithm" by Man-Hong Yung and Alán Aspuru-Guzik. *PNAS*. 2012 [RESEARCH PAPER]. https://www.pnas.org/doi/10.1073/pnas.1111758109

"A Short Introduction to Topological Quantum Computation" SciPostPhys. 2017 [RESEARCH PAPER]. https://scipost.org/SciPostPhys.3.3.021/pdf

"A Step-by-Step HHL Algorithm Walkthrough to Enhance Understanding of Critical Quantum Computing Concepts" by Anika Zaman et al. *arXiv*. 2023. https://arxiv.org/pdf/2108.09004.pdf

"Accuracy and Resource Estimations for Quantum Chemistry on a Near-term Quantum Computer" by Michael Kühn, Peter Deglmann, and Horst Weiß. *ACS*. 2018 [RESEARCH PAPER]. https://arxiv.org/pdf/1812.06814.pdf

"Aerospace: Economic Growth" by Aviation Benefits Beyond Borders Website [WEB POST]. https://aviationbenefits.org/economic-growth/

"Aerospace": https://www.britannica.com/technology/aerospace-industry

"Agriculture and Food" World Bank website. [WEB POST]. https://www.worldbank.org/en/topic/agriculture/overview

"Agriculture": https://www.nationalgeographic.org/encyclopedia/agriculture/

"Air Force Research Laboratory Partners With QC Ware to Apply Quantum Machine Learning in Identifying Flight Patterns of Unmanned Aircraft" QC Ware. 2021 [WEB POST]. https://www.qcware.com/news/air-force-research-laboratory-partners-with-qc-ware-to-apply-quantum-machine-learning-in-identifying-flight-patterns-of-unmanned-aircraft

"Airbus Gets Aerodynamic With Quantum Computing" The Next Platform. 2019 [WEB POST]. https://www.nextplatform.com/2019/01/24/airbus-gets-aerodynamic-with-quantum-computing/

"An Introduction to Resource Estimation" 2023 [WEB POST]. https://learn.microsoft.com/en-us/azure/quantum/intro-to-resource-estimation

"Analyzing the Economic Impacts of Telecommunications" Utilities One website. 2023 [WEB POST]. Utilities One web site, accessed on September 2023.

"Application Overview of Quantum Computing for Gas Turbine Design and Optimization" by Aurthur Vimalachandran Thomas Jayachandran. *IntechOpen Journals*. 2022 [RESEARCH ARTICLE]. https://www.intechopen.com/journals/1/articles/106

"Application-Oriented Performance Benchmarks for Quantum Computing" by Thomas Lubinski et al. IEEE. 2023 [RESEARCH PAPER]. https://ieeexplore.ieee.org/document/10061574

"Applying Quantum Algorithms to Satellite Mission Planning Optimization" Tech UK. 2023 [WEB POST]. https://www.techuk.org/resource/applying-quantum-algorithms-to-satellite-mission-planning-optimization-terra-quantum-and-thales-group-unlocked-new-revenue-potential.html

"Are Quantum Computers the Future of Genome Analysis?" Osaka University. 2023 [WEB POST]. https://resou.osaka-u.ac.jp/en/research/2023/20230726_1

"Are Quantum Computers the Future of Genome Analysis?" PhysOrg. 2023 [RESEARCH PAPER]. https://phys.org/news/2023-11-quantum-gene-relationships.html

"Assessing Requirements to Scale to Practical Quantum Advantage" by Michael E. Beverland. *arXiv*. 2022 [RESEARCH PAPER]. https://arxiv.org/abs/2211.07629.pdf

"Automakers Like BMW Are Becoming Quantum Computing's Early Adopters" Tech Monitor 30. 2022 [WEB POST]. https://techmonitor.ai/technology/emerging-technology/quantum-computing-automotive-bmw-pasqal

"Automotive Industry Worldwide – Statistics & Facts" Statista. [WEB POST]. https://www.statista.com/topics/1487/automotive-industry/#topicOverview

"Automotive": https://www.britannica.com/technology/automotive-industry

"Average Profit Margin for Telecommunications Agency: Overview" Investopedia. 2022 [WEB POST]. https://www.investopedia.com/ask/answers/060215/what-average-profit-margin-company-telecommunications-sector.asp

"AWS, Partners Report Successful Quantum Key Distribution Trial in Singapore" by John Russell HPC Wire. 2023 [PRESS RELEASE]. https://www.hpcwire.com/2023/03/06/aws-partners-report-successful-quantum-key-distribution-trial-in-singapore/

"Azure Quantum Elements Aims to Compress 250 Years of Chemistry into the Next 25" Microsoft Website. 2023 [WEB POST]. https://news.microsoft.com/source/features/innovation/azure-quantum-elements-chemistry-materials-science/

"Banco Santander, BBVA and CaixaBank Team Up to Tackle Fraud with FrauDfense" Fintech Futures. 2023 [WEB POST]. https://www.fintechfutures.com/2023/07/banco-santander-bbva-and-caixabank-unify-fraud-defences-with-fraudfense/

"BASF Collaborates with PASQAL to Predict Weather Patterns" Pasqal. 2022 [PRESS RELEASE]. https://www.pasqal.com/articles/basf-collaborates-with-pasqal-to-predict-weather-patterns

"Bayer Quantum-inspired Computing's Potential to Raise Yields" Fujitsu. 2022 [CASE STUDY].

"BBVA and Multiverse Showcase How Quantum Computing Could Help Optimize Investment Portfolio Management" BBVA. 2022 [WEB POST]. https://www.bbva.com/en/bbva-and-multiverse-showcase-how-quantum-computing-could-help-optimize-investment-portfolio-management/

"Beyond the Hype: A Critical Look at Quantum Computing's Potential for Business and Society in Asia-Pacific" EY Website [WHITE PAPER]. https://assets.ey.com/content/dam/ey-sites/ey-com/en_gl/topics/financial-services-asia-pacific/beyond-the-hype-a-critical-look-at-quantum-computings-potential-for-business-and-society-in-asia-pacific.pdf; https://www.eneos-innovation.co.jp/english/newsroom/20200421

"Blueprint for a Scalable Photonic Fault-Tolerant Quantum Computer" by J. Eli Bourassa, et al. *Quantum Open Journal*. 2021 [RESEARCH PAPER]. https://quantum-journal.org/papers/q-2021-02-04-392/

"BMO Financial Group and Scotiabank Partner with Xanadu on Quantum Computing Speedups for Trading Products" Xanadu. 2019 [PRESS RELEASE]. https://www.prnewswire.com/news-releases/bmo-financial-group-and-scotiabank-partner-with-xanadu-on-quantum-computing-speedups-for-trading-products-300904106.html

"BMW Developing More Durable, Safer and Less Expensive Next Generation Batteries" [CASE STUDY]. https://www.pasqal.com/industry/mobility/battery-modelling

"Bosch to Build Quantum Digital Twin of Madrid Factory" Yole Group. 2022 [WEB POST]. https://www.yolegroup.com/industry-news/bosch-to-build-quantum-digital-twin-of-madrid-factory/

"BP Joins the IBM Quantum Network to Advance Use of Quantum Computing in Energy" Green Car Congress. 2021 [WEB POST]. https://www.greencarcongress.com/2021/02/20210222-bp.html

"Brief History of Quantum Cryptography: A Personal Perspective" by Gilles Brassard. *arXiv*. 2006 [RESEARCH PAPER]. https://arxiv.org/abs/quant-ph/0604072

"Can Quantum Computers Bring an End to Corrosion?" IBM Research Website. 2023 [WEB POST]. https://research.ibm.com/blog/boeing-quantum-corrosion

"Carbon-capture Technology Could Benefit from Quantum Computing" Physics World. 2023 [WEB POST]. https://physicsworld.com/a/carbon-capture-technology-could-benefit-from-quantum-computing/

"Chemical Industry Contributes $5.7 Trillion to Global GDP and Supports 120 Million Jobs" International Council of Chemical Associations (ICCA) Website. 2019 [REPORT]. https://icca-chem.org/news/chemical-industry-contributes-5-7-trillion-to-global-gdp-and-supports-120-million-jobs-new-report-shows/

"Chemicals": https://www.britannica.com/technology/chemical-industry

"Citi and Classiq Advance Quantum Solutions for Portfolio Optimization Using Amazon Braket" by Yoram Avidan et al. *AWS Website*. 2024 [BLOG POST]. https://aws.amazon.com/blogs/quantum-computing/citi-and-classiq-advance-quantum-solutions-for-portfolio-optimization/

"Cold Atom Qubits" by Dmitry Solenov and Dmitry Mozyrsky. American Scientific Publishers. 2010 [RESEARCH PAPER]. https://arxiv.org/abs/1005.5487

"Combining the QAOA and HHL Algorithm to Achieve a Substantial Quantum Speedup for the Unit Commitment Problem" by Jonas Stein et al. 2023 [RESEARCH PAPER]. https://arxiv.org/pdf/2305.08482.pdf

"Communication Theory of Secrecy Systems" by Claude E. Shannon. *Bell System Technical Journal*, 1949 [RESEARCH PAPER]. https://ieeexplore.ieee.org/document/6769090

"Conjugate Coding" by Stephen Weisner. *ACM SIGACT*, 1983 [RESEARCH PAPER]. https://dl.acm.org/doi/10.1145/1008908.1008920

"Cooling Quantum Computers: Keeping Your Qubits Stable Requires Some of the Most Extreme Cooling Equipment Around" by Sebastian Moss. *DCD*. 2021 [WEB POST]. https://www.datacenterdynamics.com/en/analysis/cooling-quantum-computers/

"Could Quantum Computing Clean Up the Ongoing Air Travel Mess?" Inside Quantum Technology. 2022 [WEB ARTICLE]. https://www.insidequantumtechnology.com/news-archive/could-quantum-computing-clean-up-the-ongoing-air-travel-mess/;

"Could Quantum Computing Make Our Energy Grid Sustainable?" Tech Monitor. 2023 [WEB POST]. https://techmonitor.ai/hardware/quantum/can-quantum-computing-make-the-energy-grid-sustainable

"Crossing the Chasm – Geoffrey Moore" Strategies for Influence. [WEB POST]. https://strategiesforinfluence.com/crossing-the-chasm-geoffrey-moore/

"Delta Partners with IBM to Explore Quantum Computing – An Airline Industry First" IBM. 2020 [WEB POST]. https://newsroom.ibm.com/2020-01-08-Delta-Partners-with-IBM-to-Explore-Quantum-Computing-an-Airline-Industry-First

"Demystifying Cybersecurity: Post-Quantum Security" ISEC7. 2023 [WEB POST].
https://blog.isec7.com/en/demystifying-cybersecurity-post-quantum-security-part-1
https://blog.isec7.com/en/demystifying-cybersecurity-post-quantum-security-part-2

"Digital Technologies to Achieve the UN SDGs" by ITU. 2021 [WEB POST]. https://www.itu.int/en/mediacentre/backgrounders/Pages/icts-to-achieve-the-united-nations-sustainable-development-goals.aspx

"Digitized Counterdiabatic Quantum Algorithm for Protein Folding" by APS. 2023 [RESEARCH PAPER]. https://journals.aps.org/prapplied/abstract/10.1103/PhysRevApplied.20.014024

"DLR and NASA Are Jointly Developing a Software Package for Quantum Computers" DLR. 2022 [WEB POST]. https://www.dlr.de/en/latest/news/2022/01/20220303_dlr-and-nasa-developing-software-package-for-quantum-computers

"E.ON to Manage Decentralised Energy System Using Quantum Computing" Smart Energy International. 2021 [WEB POST]. https://www.smart-energy.com/industry-sectors/digitalisation/e-on-to-manage-decentralised-energy-system-using-quantum-computing/

"Efficient Quantum Circuits for Schur and Clebsch-Gordan Transforms," by D. Bacon, I. L. Chuang, and A. W. Harrow, *Phys. Rev. Lett.* 97, 170502 (2006); "Efficient Compression of Quantum Information," by M. Plesch and V. Buzek, *Phys. Rev. A* 81, 032317 (2010).

"Electronics": https://www.investopedia.com/ask/answers/042915/what-electronics-sector.asp

"Elliptic Curve Cryptography: What Is It? How Does It Work? Keyfactor [WEB POST]. https://www.keyfactor.com/blog/elliptic-curve-cryptography-what-is-it-how-does-it-work/

"Elucidating Reaction Mechanisms on Quantum Computers" by Markus Reiher et al. *PNAS.* 2017 [RESEARCH PAPER]. https://www.pnas.org/doi/10.1073/pnas.1619152114

"ENEOS Leads the Way to Sustainable Hydrogen Fuel by Validating Computing Algorithms Using Azure Quantum" Microsoft. 2022 [CASE STUDY]. https://customers.microsoft.com/en-us/story/1508526643598641548-eneos-energy-azure-quantum

"Energy": https://www.investopedia.com/terms/e/energy_sector.asp

"Eni and PASQAL Together to Develop Quantum Solutions for the Energy Sector" Eni. 2022 [USE CASE]. https://www.eni.com/en-IT/media/press-release/2022/11/eni-pasqal-together-develop-quantum-solutions-energy-sector.html

"Entanglement and Quantum Information", by P.G. Kwiat, D.F.V. James. Science Direct. 2023 [WEB POST]. https://www.sciencedirect.com/topics/engineering/quantum-teleportation

"Error Mitigation Extends the Computational Reach of a Noisy Quantum Processor" by Abhinav Kandala et al. *Nature.* 2019. https://www.nature.com/articles/s41586-019-1040-7

"Explainer: What Is Quantum Communication?" by Martin Giles. *MIT Technology Review.* 2019 [WEB POST]. https://www.technologyreview.com/2019/02/14/103409/what-is-quantum-communications/

"Exploiting Fermion Number in Factorized Decompositions of the Electronic Structure Hamiltonian" by Sam McArdle. 2022 [RESEARCH PAPER]. https://arxiv.org/pdf/2107.07238.pdf

"Explore Quantum. Quantum Error Correction" Quantum Microsoft Web. [WEB POST]. https://quantum.microsoft.com/en-us/explore/concepts/quantum-error-correction#:~:text=Current%20state%2Dof%2Dthe%2D,a%201%20or%20vice%20versa

"Exponential Algorithmic Speedup by Quantum Walk" by Andrew M. Childs et al. *arXiv.* 2002 [RESEARCH PAPER]. https://arxiv.org/pdf/quant-ph/0209131.pdf

"ExxonMobil & IBM Explore Quantum Algorithms to Solve Routing Formulations" IBM Medium. 2021 [WEB POST]. https://ibm-research.medium.com/exxonmobil-ibm-scientists-explore-state-of-art-quantum-algorithms-to-solve-routing-formulations-e7ce39f8741c

"ExxonMobil, NCSA, Cray Scale Reservoir Simulation to 700,000+ Processors" HPCwire. 2017 [PRESS RELEASE]. https://www.hpcwire.com/2017/02/17/exxonmobil-ncsa-cray-scale-reservoir-simulation-700000-processors/

"Fault-tolerant Resource Estimate for Quantum Chemical Simulations: Case study on Li-ion Battery Electrolyte Molecules" by Isaac H. Kim. *APS.* 2022 [RESEARCH PAPER]. https://journals.aps.org/prresearch/abstract/10.1103/PhysRevResearch.4.023019

"FedEx's Investments in Quantum Computing: What's on the Horizon" Techbuillon. 2023 [WEB ARTICLE]. https://techbullion.com/fedexs-investments-in-quantum-computing-whats-on-the-horizon/

"Fertilizer and Other Quantum Computer Chemistry" Quantum Flagship. [WEB POST]. https://qt.eu/applications/fertilizer-and-other-quantum-computer-chemistry

"Finance": https://www.investopedia.com/terms/f/financial_sector.asp

"Financial Services: Sizing the Sector in the Global Economy" Investopedia. 2021 [WEB POST]. https://www.investopedia.com/ask/answers/030515/ what-percentage-global-economy-comprised-financial-services-sector.asp

"Finding flows of a Navier–Stokes Fluid through Quantum Computing" by Frank Gaitan. *Nature*. 2020 [RESEARCH PAPER]. https://www.nature.com/ articles/s41534-020-00291-0

"Ford Enlists Quantum Computing in EV Battery Materials Hunt" EE Times Europe. 2023 [WEB POST]. https://www.eetimes.eu/ford-enlists-quantum-computing-in-ev-battery-materials-hunt/

"Ford, Microsoft Partner to Reduce Congestion with 'Quantum-inspired' Tech" Ford Media. 2019 [PRESS RELEASE]. https://media.ford.com/ content/fordmedia/fna/us/en/news/2019/12/10/ford-exploring-quan-tum-world-with-microsoft.html

"Formulating and Solving Routing Problems on Quantum Computers" by Stuart Harwood et al. IEEE. 2021 [RESEARCH PAPER]. https://ieeexplore.ieee. org/document/9314905

"Fuel Consumption Minimization of Transport Aircraft Using Real-Coded Genetic Algorithm" *Sage Journals*. 2017 [RESEARCH PAPER]. https://journals.sage-pub.com/doi/full/10.1177/0954410017705899

"Global Pharmaceutical Industry – Statistics & Facts" Statista Key Insights [WEB POST]. https://www.statista.com/topics/1764/global-pharmaceutical-industry/#topicOverview

"Global Supply Chain Management (SCM) Market to Grow from $28.9 Billion to $45.2 Billion by 2027 at a CAGR of 9.4%" Research and Markets. 2022 [PRES RELEASE]. https://www.globenewswire.com/en/news-release/2022/ 06/29/2471248/28124/en/Global-Supply-Chain-Management-SCM-Market-to-Grow-from-28-9-Billion-to-45-2-Billion-by-2027-at-a-CAGR-of-9-4.html

"Government": https://www.britannica.com/summary/government

"GSK on Quantum Use Cases in Pharmaceuticals" IoT World Today. 2022 [VIDEO]. https://www.quantumbusinessnews.com/applications/video-gsk-on-quan-tum-computing-use-cases-for-the-pharmaceutical-industry

"How BMW Can Maximize Its Supply Chain Efficiency with Quantum" Honeywell. [News Post]. https://www.honeywell.com/us/en/news/2021/01/ exploring-supply-chain-solutions-with-quantum-computing

"How Many People Work for the Federal Government?" USA Facts. 2023 [WEB POST]. https://usafacts.org/articles/how-many-people-work-for-the-fed-eral-government/#:~:text=As%20of%20September%202023%2C%20 there,increasing%20again%20in%20the%201950s.

"How Much Revenue Has the U.S. Government Collected This Year?" FiscalData. [WEB POST]. https://fiscaldata.treasury.gov/americas-finance-guide/government-revenue/

"How NASA Utilizes Quantum Computing for Weather Forecasting" Tech Buillon. 2023 [WEB POST]. https://techbullion.com/how-nasa-utilizes-quantum-computing-for-weather-forecasting/

"How Quantum Computers Could Help Design Airplanes" IBM. 2023 [CASE STUDY]. https://research.ibm.com/blog/boeing-case-study

"How to Factor 2048 Bit RSA Integers in 8 Hours Using 20 Million Noisy Qubits" by Craig Gidney and Martin Ekerå. *arXiv*. 2019 [RESEARCH PAPER]. https://arxiv.org/abs/1905.09749

"How VW, Bosch, Ford, Daimler Aim to Gain from Quantum Computing" Europe Autonews. 2019 [WEB POST]. https://europe.autonews.com/automakers/how-vw-bosch-ford-daimler-aim-gain-quantum-computing

"Hybrid Quantum-classical Algorithms in the Noisy Intermediate-scale Quantum Era and beyond" by Adam Callison and Nicholas Chancellor. *Phys. Rev.* 2022 [RESEARCH PAPER]. https://journals.aps.org/pra/abstract/10.1103/PhysRevA.106.010101

"Industry vs. Sector: What's the Difference?": https://www.investopedia.com/ask/answers/05/industrysector.asp

"Inside PayPal's Partnership with IBM to Use Quantum Computing to Improve How It Detects Fraud and Underwrites" Business Insider. 2022 [WEB POST]. https://www.businessinsider.com/paypal-quantum-computing-fraud-prevention-lending-machine-learning-payments-tech-2022-1

"IonQ and Hyundai Motor Partner to Use Quantum Computing to Advance Effectiveness of Next-Gen Batteries" Hyundai. 2022 [PRESS RELEASE]. https://www.hyundai.news/eu/articles/press-releases/ionq-and-hyundai-partner-to-use-quantum-computing-to-advance-effectiveness-of-next-gen-batteries.html

"IonQ, Airbus Sign Agreement to Collaborate on Aircraft Loading Project Using Quantum Computing" IonQ Web. 2022 [WEB POST]. https://ionq.com/news/august-18-2022-ionq-2022-airbus

"Jülich Quantum Computer Solves Protein Puzzle" Julich. 2023 [WEB POST]. https://www.fz-juelich.de/en/news/archive/press-release/2023/julich-quantum-computer-solves-protein-puzzle

"Linear Algebra for Quantum Computing" Microsoft. 2023 [WEB POST]. https://learn.microsoft.com/en-us/azure/quantum/overview-algebra-for-quantum-computing#vectors-and-matrices-in-quantum-computing

"Low Depth Amplitude Estimation on a Trapped Ion Quantum Computer" by Tudor Giurgica-Tiron. *arXiv*. 2021 [RESEARCH PAPER]. https://arxiv.org/abs/2109.09685

"Medical Image Classification via Quantum Neural Networks" by Natansh Mathur. *arXiv*, 2021 [RESEARCH PAPER]. https://arxiv.org/abs/2109.01831

"Mercedes-Benz Bets on Quantum to Craft the Future of Electric Vehicles" IBM. [CASE STUDY]. https://www.ibm.com/case-studies/daimler

"Middleware for Quantum: An Orchestration of Hybrid Quantum-Classical Systems" IEEE. 2023 [CONFERENCE PAPER]. https://ieeexplore.ieee.org/document/10234290

"Migration to Post-Quantum Cryptography" NIST. 2023 [PRESENTATION]. https://www.nccoe.nist.gov/crypto-agility-considerations-migrating-post-quantum-cryptographic-algorithms

"Modelling Carbon Capture on Metal-Organic Frameworks with Quantum Computing" Springer. 2022 [RESEARCH PAPER]. https://epjquantumtechnology.springeropen.com/articles/10.1140/epjqt/s40507-022-00155-w

"Moving the World's Energy Products across the Globe Is a Complex Puzzle That Could Benefit from a Quantum Solution" IBM website. 2020 [CASE STUDY]. https://www.ibm.com/case-studies/exxonmobil

"NIST Researchers Help Design a Prototype Quantum Computer" by NIST. 2023 [WEB POST]. https://www.nist.gov/news-events/news/2023/12/nist-researchers-help-design-prototype-quantum-computer

"NIST to Standardize Encryption Algorithms That Can Resist Attack by Quantum Computers" NIST. 2023 [WEB POST]. https://www.nist.gov/news-events/news/2023/08/nist-standardize-encryption-algorithms-can-resist-attack-quantum-computers

"Nomura Launches Joint Research on Quantum Computing" Nomura. 2018 [PRESS RELEASE]. https://www.nomuraholdings.com/news/nr/holdings/20180227/20180227.pdf

"Nvidia Spools Up Quantum Jet-Engine Simulations" by Edd Gent. IEEE. 2023 [RESEARCH PAPER]. https://spectrum.ieee.org/computational-fluid-dynamics-quantum-computer

"NVIDIA, Rolls-Royce and Classiq Announce Quantum Computing Breakthrough for Computational Fluid Dynamics in Jet Engines" Nvidia. 2023 [PRESS RELEASE]. https://nvidianews.nvidia.com/news/nvidia-rolls-royce-and-classiq-announce-quantum-computing-breakthrough-for-computational-fluid-dynamics-in-jet-engines

"Oil and Gas Industry Faces Moment of Truth – and Opportunity to Adapt – as Clean Energy Transitions Advance" International Energy Agency: IEA. 2023 [WEB POST]. https://www.iea.org/news/oil-and-gas-industry-faces-moment-of-truth-and-opportunity-to-adapt-as-clean-energy-transitions-advance

"OKI Uses D-Wave Quantum Computer to Increase Manufacturing Productivity" LaserFocus. 2019 [WEB POST]. https://www.laserfocusworld.com/test-measurement/article/14067894/oki-uses-dwave-quantum-computer-to-increase-manufacturing-productivity

"Optimization of Manufacturing Equipment Layouts in Factories Employing Quantum Computer" Oki. 2020 [WHITE PAPER]. https://www.oki.com/en/otr/2020/n235/pdf/otr-235-R13.pdf

"Option Pricing Using Quantum Computers" by Nikitas Stamatopoulos. *Quantum*. 2020 [RESEARCH PAPER]. https://arxiv.org/pdf/1905.02666.pdf

"Oxford Joins Consortium to Advance Quantum Drug Discovery with £6.8M Grant from Innovate UK" Oxford University. 2021 [WEB POST]. https://www.ox.ac.uk/news/2021-11-05-oxford-joins-consortium-advance-quantum-drug-discovery-68m-grant-innovate-uk

"Pharma Giant Roche Partners for Quantum Drug Discovery" The Next Platform. 2021 [PRESS RELEASE]. https://www.nextplatform.com/2021/01/28/pharma-giant-roche-partners-for-quantum-drug-discovery/

"Pharma": https://www.investopedia.com/articles/markets/051316/industry-handbook-pharma-industry.asp

"Picturing the Future with Quantum-Enabled Road Sign Recognition" IonQ. 2023 [CASE STUDY]. https://ionq.com/resources/picturing-the-future-quantum-enabled-road-sign-recognition

"Polynomial-Time Algorithms for Prime Factorization and Discrete Logarithms on a Quantum Computer" ACM. 1997 Peter Shor [RESEARCH PAPER]. https://dl.acm.org/doi/10.1137/S0097539795293172

"Practical Quantum Security for Grid Automation: Minimize Cost and Increase Security by Enabling a Grid-compatible, Long-term QKD Solution for Guaranteed Secure Communications" by US Department of Energy. 2013 [RESEARCH PROJECT]. https://www.energy.gov/sites/prod/files/2017/04/f34/ORNL_Practical_Quantum_Security_FactSheet_0.pdf

"Proof of Concept Showed Ability to Detect and Defend against Potential Threats and Eavesdroppers" by Toshiba. 2022 [PRESS RELEASE]. https://news.toshiba.com/press-releases/press-release-details/2022/JPMorgan-Chase-Toshiba-and-Ciena-Build-the-First-Quantum-Key-Distribution-Network-Used-to-Secure-Mission-Critical-Blockchain-Application/default.aspx

"Protein Folding Takes a Step Forward With Quantum Computing" QuantumZeitgeist. 2022 [WEB POST]. https://quantumzeitgeist.com/protein-folding-takes-a-step-forward-with-quantum-computing/

"Protocol to Identify a Topological Superconducting Phase in a Three-terminal Device" by Dmitry I. Pikulin et al. *arXiv*. 2021 [RESEARCH PAPER].

"Quandela and EDF Work Together to Use Photonic Quantum Computing to Simulate Hydroelectric Dam Structures" Quandela. 2022 [PRESS RELEASE]. https://www.quandela.com/wp-content/uploads/2022/11/Quandela-and-EDF-collaborate-Press-Release.pdf

"Quantum Algorithm for Linear Systems of Equations" by Aram W. Harrow, Avinatan Hassidim, and Seth Lloyd. *APS*. 2008 [RESEARCH PAPER]. https://journals.aps.org/prl/abstract/10.1103/PhysRevLett.103.150502

"Quantum Algorithms Applied to Satellite Mission Planning for Earth Observation" by Sam McArdle. *IEEE Journal of Selected Topics in Applied Earth Observations and Remote Sensing*. 2023 [RESEARCH PAPER]. https://ieeexplore.ieee.org/stamp/stamp.jsp?arnumber=10155128

"Quantum Algorithms: A Survey of Applications and End-to-end Complexities" by Alexander M. Dalzell et al. *arXiv*. 2023 [RESEARCH PAPER]. https://arxiv.org/pdf/2310.03011.pdf

"Quantum Amplitude Amplification and Estimation" by Gilles Brassard et al. *arXiv*. 2000 [RESEARCH PAPER]. https://arxiv.org/pdf/quant-ph/0005055.pdf

"Quantum and Classical Machine Learning for the Classification of Non-small-cell Lung Cancer Patients" by Siddhant Jain et al. 2020 [RESEARCH PAPER]. https://link.springer.com/article/10.1007/s42452-020-2847-4

"Quantum Chemistry in the Age of Quantum Computing" by Yudong Cao et al. *ACS*, 2019 [RESEARCH PAPER]. https://pubs.acs.org/doi/10.1021/acs.chemrev.8b00803

"Quantum Communications to Move towards an Ultra-secure Smart Grid" by TECNALIA. 2023 [PRESS RELEASE]. https://www.tecnalia.com/en/press-room/quantum-communications-to-move-towards-an-ultra-secure-smart-grid

"Quantum Complexity Theory" by Ethan Bernstein and Umesh Vazirani. *ACM*, 1993 [RESEARCH PAPER]. https://dl.acm.org/doi/10.1145/167088.167097

"Quantum Computations with Cold Trapped Ions" by J. I. Cirac and P. Zoller. *Phys. Rev.*, 1995 [RESEARCH PAPER]. https://iontrap.umd.edu/wp-content/uploads/2013/10/Quantum-computations-with-cold-trapped-ions.pdf

"Quantum Computers for Weather and Climate Prediction: The Good, the Bad and the Noisy" American Meteorological Society. 2023 [RESEARCH PAPER]. https://arxiv.org/abs/2210.17460

"Quantum Computers Take Us to New Dimensions" Infineon. [WEB POST]. https://www.infineon.com/cms/en/product/promopages/quantumcomputing/

"Quantum Computing and Genome Sequencing Unlocking Genetic Mysteries" UtiliesOne. 2023 [WEB POST].

"Quantum Computing and Materials Science: A Practical Guide to Applying Quantum Annealing to the Configurational Analysis of Materials" by B. Camino. *Appl. Phys.*, 2023 [RESEARCH ARTICLE]. https://pubs.aip.org/aip/jap/article/133/22/221102/2896017/Quantum-computing-and-materials-science-A

"Quantum Computing and Simulations for Energy Applications" by Hari Paudel. IEEE International Conference on Quantum Computing and Engineering QCE22. 2022 [PRESENTATION]. https://www.osti.gov/servlets/purl/1971602

"Quantum Computing Breakthrough Heralds a New Era of Jet Engine Design" Financial Express. 2023 [WEB POST]. https://www.financialexpress.com/business/defence-a-quantum-computing-breakthrough-heralds-a-new-era-of-jet-engines-3119729/

"Quantum Computing Breakthrough in Simulating Chemical Molecules" 2023 [WEB POST]. https://www.eetimes.eu/quantum-computing-breakthrough-in-simulating-chemical-molecules/

"Quantum Computing Enables Unprecedented Materials Science Simulations" Office of Science. 2021 [WEB POST]. https://www.energy.gov/science/ascr/articles/quantum-computing-enables-unprecedented-materials-science-simulations

"Quantum Computing for Chemical and Biomolecular Product Design" by Martin P. Andersson et al. Science Direct. 2022 [RESEARCH PAPER]. https://www.sciencedirect.com/science/article/pii/S2211339821000861

"Quantum Computing for Chemistry" Kassal Group. [WEB POST]. https://www.kassal.group/research/quantum/

"Quantum Computing for Energy Systems Optimization: Challenges and Opportunities" by Akshay Ajagekar, Fengqi You. *arXiv*. 2020 [RESEARCH PAPER]. https://arxiv.org/pdf/2003.00254

"Quantum Computing Governance Principles" by WEF. 2022 [INSIGHT REPORT]. https://www3.weforum.org/docs/WEF_Quantum_Computing_2022.pdf

"Quantum Computing in Material Discovery" Infosys. [WEB POST]. https://blogs.infosys.com/emerging-technology-solutions/quantum-computing/quantum-computing-in-material-discovery.html

"Quantum Computing is Becoming Business Ready" BCG. 2023 [REPORT]. https://www.bcg.com/publications/2018/next-decade-quantum-computing-how-play

"Quantum Computing May Create Ethical Risks for Businesses. It's Time to Prepare" by a Scott Buchholz and Beena Ammanath. *Deloitte Insights*. 2022 [WEB POST]. https://www2.deloitte.com/us/en/insights/topics/cyber-risk/quantum-computing-ethics-risks.html

"Quantum Computing on Azure | How It Works, What's Coming, & What You Can Try Today" Microsoft Blog. 2023 [WEB POST]. https://techcommunity.microsoft.com/t5/microsoft-mechanics-blog/quantum-computing-on-azure-how-it-works-what-s-coming-amp-what/ba-p/3762127

"Quantum Computing: A New Threat to Cybersecurity" by Michele Mosca. Global Risk Institute. for Quantum Computing [REPORT]

"Quantum Computing: Boehringer Ingelheim and Google Partner for Pharma R&D" Boehringer-Ingelheim website. 2021 [News Post]. https://www.boehringer-ingelheim.com/press-release/partnering-google-quantum-computing

"Quantum Computing" Rolls Royce. 2023 [WEB POST]. https://www.rolls-royce.com/media/our-stories/discover/2023/trent-xwb-quantum-computing.aspx

"Quantum Data Encoding: A Comparative Analysis of Classical-to-Quantum Mapping Techniques and Their Impact on Machine Learning Accuracy" by Minati Rath and Hema Date. 2023 [RESEARCH PAPER]. https://arxiv.org/pdf/2311.10375.pdf

"Quantum Dots: Everything You Need to Know" EDN. 2019 [WEB POST]. https://www.edn.com/quantum-dots-explained/#google_vignette

"Quantum Encoding: An Overview" Quantum Zeitgeist. 2021 [WEB POST]. https://quantumzeitgeist.com/quantum-encoding-an-overview/

"Quantum Encryption vs. Post-Quantum Cryptography" QuantumXC. 22 March 2022. https://quantumxc.com/blog/ quantum-encryption-vs-post-quantum-cryptography-infographic/

"Quantum Machine Learning Beyond Kernel Methods" by Sofiene Jerbi et al. *Nature*. 2023 [RESEARCH PAPER]. https://www.nature.com/articles/s41467-023-36159-y

"Quantum Machine Learning for Better Image Recognition" QM Ware. 2023 [WEB POST]. https://www.qm-ware.com/use-cases/quantum-machine-learning-automotive/

"Quantum Machine Learning: Benefits and Practical Examples" by Frank Phillipson. *CEUR Workshop*. 2020 [RESEARCH PAPER]. https://ceur-ws.org/Vol-2561/paper5.pdf

"Quantum Machine Learning" by Jacob Biamonte. *Nature*. 2017 [RESEARCH PAPER]. https://www.nature.com/articles/nature23474

"Quantum Neuromorphic Computing" by Danijela Marković and Julie Grollier. *arXiv*. 2020 [RESEARCH PAPER]. https://arxiv.org/abs/2006.15111

"Quantum Pricing with a Smile: Implementation of Local Volatility Model on Quantum Computer" by Kazuya Kaneko. Springer. 2022 [RESEARCH PAPER]. https://epjquantumtechnology.springeropen.com/articles/10.1140/epjqt/s40507-022-00125-2

"Quantum Random Number Generation" by Xiongfeng Ma et al. *Nature*. 2016 [RESEARCH PAPER]. https://www.nature.com/articles/npjqi201621

"Quantum Random Number Generators" by Miguel Herrero-Collantes and Juan Carlos Garcia-Escartin. *arXiv*. 2016 [RESEARCH PAPER]. https://arxiv.org/pdf/1604.03304.pdf

"Quantum Speedup of Monte Carlo Methods" by Ashley Montanaro. *arXiv*. 2017 [RESEARCH PAPER]. https://arxiv.org/pdf/1504.06987.pdf

"Quantum Speedup of the Travelling Salesman Problem for Bounded-degree Graphs" by Alexandra E. Moylett et al. *arXiv*. 2016 [RESEARCH PAPER]. https://arxiv.org/pdf/1612.06203.pdf

"Quantum Support Vector Machines for Aerodynamic Classification" by Xi-Jun Yuan et al. *Intelligent Computing*. 2023 [RESEARCH PAPER]. https://spj.science.org/doi/pdf/10.34133/icomputing.0057

"Quantum Technologies: A Potential Game-changer in Aerospace" Airbus. [WEB POST]. https://www.airbus.com/en/innovation/disruptive-concepts/quantum-technologies

"Quantum Technology Monitor" by McKinsey. 2023 [REPORT]. https://www.mckinsey.com/~/media/mckinsey/business%20functions/mckinsey%20digital/our%20insights/quantum%20technology%20sees%20record%20investments%20progress%20on%20talent%20gap/quantum-technology-monitor-april-2023.pdf

"Quantum Technology Strengthens Medical Imaging Methods through Quanvolutional Neural Network" Quantum Zeitgeist. 2022 [WEB POST]. https://quantumzeitgeist.com/quantum-technology-strengthens-medical-imaging-methods-through-quanvolutional-neural-network/

"Quantum Technology: The Second Quantum Revolution" by Jonathan P. Dowling and Gerard J. Milburn. The Royal Society Publishing. 2003 [RESEARCH PAPER]. https://royalsocietypublishing.org/doi/epdf/10.1098/rsta.2003.1227

"Quantum-readiness: Migration to Post-quantum Cryptography" by NIST. 2023 [FACT SHEET]. https://www.nccoe.nist.gov/sites/default/files/2023-08/quantum-readiness-fact-sheet.pdf

"Quantum-Safe Cryptography (QSC)" ETSI. [WEB POST]. https://www.etsi.org/technologies/quantum-safe-cryptography#:~:text=The%20ETSI%20Cyber%20Quantum%20Safe,algorithm%20research%2C%20as%20well%20as

"Qubit Engineering Optimizes Wind Farm Energy Production with Azure Quantum" Microsoft. 2022 [WEB POST]. https://cloudblogs.microsoft.com/quantum/2022/05/17/qubit-engineering-optimizes-wind-farm-energy-production-with-azure-quantum/

"Qubit Engineering Optimizes Wind Farm Energy Production with Azure Quantum" Microsoft. 2022 [WEB POST]. https://cloudblogs.microsoft.com/quantum/2022/05/17/qubit-engineering-optimizes-wind-farm-energy-production-with-azure-quantum/

"Rapid Solutions of Problems by Quantum Computation" by David Deutsch and Richard Jozsa. Proceedings of the Royal Society of London. 1992 [RESEARCH PAPER]. https://royalsocietypublishing.org/doi/10.1098/rspa.1992.0167

"Reducing the Number of Qubits from $n2$ to $n \log2 (n)$ to Solve the Traveling Salesman Problem with Quantum Computers: A Proposal for Demonstrating Quantum Supremacy in the NISQ Era" by Mehdi Ramezani et al. *arXiv*. 2024 [RESEARCH PAPER]. https://arxiv.org/pdf/2402.18530.pdf

"Resource-Efficient Quantum Algorithm for Protein Folding" by Anton Robert et al. *Nature*. 2021 [RESEARCH PAPER]. https://www.nature.com/articles/s41534-021-00368-4

"Rigetti Enhances Predictive Weather Modeling with Quantum Machine Learning" Rigetti Computing. 2021 [PRESS RELEASE]. https://www.globenewswire.com/news-release/2021/12/01/2344216/0/en/Rigetti-Enhances-Predictive-Weather-Modeling-with-Quantum-Machine-Learning.html

"Robotics and Computer-Integrated Manufacturing" by Marcelo Luis Ruiz Rodríguez. *ScienceDirect*. 2022 [RESEARCH PAPER]. https://www.science-direct.com/science/article/pii/S0736584522000928

"Roche Quantum Neural Network" QCware. 2021 [CASE STUDY]. https://www.qcware.com/customers/roche

"Rolls-Royce, Riverlane Partner on Quantum Materials Simulation" IoT World Today. 2023 [WEB POST]. https://www.iotworldtoday.com/quantum/rolls-royce-riverlane-partner-on-quantum-materials-simulation

"Secure Quantum Key Distribution with Realistic Devices" by Feihu Xu et al. *APS*. 2020 [RESEARCH PAPER]. https://journals.aps.org/rmp/abstract/10.1103/RevModPhys.92.025002

"Simulating a Discrete Time Crystal" Medium Qiskit. 2021 [WEB POST]. https://medium.com/qiskit/simulating-a-discrete-time-crystal-using-qiskit-531bda6663b7

"Smart Agriculture Decision Making Scheme Using Quantum Annealing" IEEE International Conference on Quantum Computing and Engineering, 2022 [RESEARCH PAPER]. https://ieeexplore.ieee.org/document/9951228

"Steady Progress in Approaching the Quantum Advantage." 2024 [REPORT]. https://www.mckinsey.com/capabilities/mckinsey-digital/our-insights/steady-progress-in-approaching-the-quantum-advantage

"Superconducting Quantum Bits" by John Clarke and Frank K. Wilhelm. *Nature*. 2008 [RESEARCH PAPER]. https://www.nature.com/articles/nature07128

"Supply Chain": https://www.investopedia.com/terms/l/logistics.asp

"Sustainable Development. SDG 17 goals" by UN. [WEB POST]. https://sdgs.un.org/goals

"Technology Quarterly. Here, There and Everywhere" The Economist. [NEWS POST]. https://www.economist.com/news/essays/21717782-quantum-technology-beginning-come-its-own

"Technology Readiness Levels" NASA. 2023 [WEB POST]. https://www.nasa.gov/directorates/somd/space-communications-navigation-program/technology-readiness-levels/

"Telecommunications": https://www.investopedia.com/ask/answers/070815/what-telecommunications-sector.asp

"Teleporting an Unknown Quantum State via Dual Classical and Einstein-Podolsky-Rosen Channels" by Charles Benett et al. *Phys. Rev. Lett.* 1993 [RESEARCH PAPER].

"The Case for Quantum and Quantum-Inspired Computing in Financial Services" Fujistu. [WHITE PAPER]. https://www.fujitsu.com/global/digitalannealer/pdf/wp-da-financialsector-ww-en.pdf

"The Diamond Quantum Revolution" Physics World. 2020 [WEB POST]. https://physicsworld.com/a/the-diamond-quantum%E2%80%AFrevolution/

"The Next Decade in Quantum Computing—and How to Play" by BCG. 2018 [REPORT]. https://www.bcg.com/publications/2018/next-decade-quantum-computing-how-play

"The Pharmaceutical Industry in Figures" European Federation of Pharmaceutical Industries and Associations. 2023 [WEB POST]. https://www.efpia.eu/media/637143/the-pharmaceutical-industry-in-figures-2022.pdf

"The Physical Implementation of Quantum Computation" by David P. DiVincenzo. *arXiv.* 2000 [RESEARCH PAPER]. https://arxiv.org/abs/quant-ph/0002077

"The Quantum Revolution: Networking and Security for Tomorrow's Internet" by CISCO. 2023 [WEB POST]. https://newsroom.cisco.com/c/r/newsroom/en/us/a/y2023/m06/the-quantum-revolution-networking-and-security-for-tomorrows-internet.html

"The Quantum Tortoise and the Classical Hare: A Simple Framework for Understanding Which Problems Quantum Computing Will Accelerate (And Which It Will Not)" by Sukwoong Choi, William S. Moses and Neil Thompson. *arXiv.* 2023 [RESEARCH PAPER]. https://arxiv.org/abs/2310.15505

"The Quest for the Ultimate Seed Begins With Quantum Computing" Global AgInvesting. 2021 [WEB POST]. https://www.globalaginvesting.com/contributed-content-quest-ultimate-seed-begins-quantum-computing/

"The Race toward Quantum Advantage" Semiconductor Engineering. 2023 [WEB POST]. https://semiengineering.com/the-race-toward-quantum-advantage/

"The Second Quantum Revolution" NIST. 2018 [WEB POST]. https://www.nist.gov/physics/introduction-new-quantum-revolution/second-quantum-revolution

"Thinking through the Ethics of New Tech···Before There's a Problem" by Beena Ammanath. *Harvard Business Review Publishing.* 2021 [WEB ARTICLE]. https://hbr.org/2021/11/thinking-through-the-ethics-of-new-techbefore-theres-a-problem

"Total is Exploring Quantum Algorithms to Improve CO2 Capture" Total Energies. 2020 [WEB POST]. https://totalenergies.com/media/news/news/total-exploring-quantum-algorithms-improve-co2-capture

"Towards the Industrialisation of Quantum Key Distribution in Communication Networks: A Short Survey" by Ruiqi Liu et al. Wiley. 2022 [RESEARCH PAPER].

"Toyota Central R&D Labs: A Quantum Approach to Transportation" D-Wave. 2021 [CASE STUDY]. https://www.dwavesys.com/media/b4lllchm/dwave_toyota_case_story_v4.pdf

"Toyota Central R&D Labs: A Quantum Approach to Transportation" DWave. 2021 [CASE STUDY]. https://www.dwavesys.com/media/b4lllchm/dwave_toyota_case_story_v4.pdf

"Toyota Harnesses Quantum Computers to Develop battery Materials" Asia Nikkei. 2021 [PRESS RELEASE]. https://asia.nikkei.com/Business/Technology/Toyota-harnesses-quantum-computers-to-develop-battery-materials

"Toyota Quantum Annealing of Vehicle Routing Problem with Time, State and Capacity" by Hirotaka Irie et al. *arXiv*. 2019 [RESEARCH PAPER]. https://arxiv.org/pdf/1903.06322.pdf

"US Defense Spending Compared to Other Countries" Peter G. Foundation. 2023 [WEB POST]. https://www.pgpf.org/chart-archive/0053_defense-comparison

"US Navy, Air Force Collaborate on Quantum" IoT World Day. 2023 [WEB POST]. https://www.quantumbusinessnews.com/applications/us-navy-air-force-collaborate-on-quantum

"Using Quantum Computing to Tell the Weather" Azo Quantum. 2018 [WEB POST]. https://www.azoquantum.com/Article.aspx?ArticleID=98#:~:text=Quantum%20Computing%20Improves%20Weather%20Forecasting&text=It%20employs%20IBM's%20supercomputing%20technology,supercomputers%20are%20unable%20to%20achieve

"Using Zapata's Quantum Workflows Platform, Orquestra, KAUST Will Explore How Quantum Computing Can Simulate and Optimize the Aerodynamic Design Process for Vehicles" Zapata. [WEB POST]. https://zapata.ai/news/zapata-partner-with-kaust-to-bring-quantum-computing-to-the-middle-east-for-the-advancement-of-computational-fluid-dynamics/

"Variational Quantum Algorithms" by M. Cerezo et al. *Nature*. 2021 [RESEARCH PAPER]. https://www.nature.com/articles/s42254-021-00348-9, https://arxiv.org/pdf/2012.09265.pdf

"Volkswagen Demonstrates First Successful Real-world Use of Quantum Computing to Help Optimize Traffic Routing" Green Car Congress. 2019 [WEB POST]. https://www.greencarcongress.com/2019/12/20191206-vwquantum.html

"Volkswagen Has Developed a Quantum-inspired Algorithm for Automotive Image Recognition That Is More Effective Than Classical Machine Learning Neural Networks" Eenews. 2023 [WEB POST]. https://www.eenewseurope.com/en/quantum-algorithm-boost-for-automotive-image-sensors/

"Volkswagen: Navigating Tough Automotive Tasks with Quantum Computing" Dwave. 2021 [CASE STUDY]. https://www.dwavesys.com/media/2pojgtcx/dwave_vw_case_story_v2f.pdf

"What Does the Future of Worker-Management Dialogue Look Like?" The Threads. [WEB POST]. https://thethreads.org/contribution/what-does-the-future-of-worker-management-dialogue-look-like/

"What Is an Economic Sector and How Do the 4 Main Types Work?": https://www.investopedia.com/terms/s/sector.asp

"What Is Cryptographic Agility? How to Get Crypto-Agility?" Encryption Consulting. [WEB POST]. https://www.encryptionconsulting.com/education-center/what-is-crypto-agility

"What Is Hilbert Space?" [WEB POST]. https://www.scienceabc.com/pure-sciences/what-is-a-hilbert-space.html

"What Is Quantum Annealing?" D-WAVE. [WEB POST]. https://docs.dwavesys.com/docs/latest/c_gs_2.html

"What Is Quantum Processing Unit (QPU)" Huawei. 2023 [WEB POST]. https://forum.huawei.com/enterprise/en/what-is-quantum-processing-unit-qpu/thread/667282964481982464-667213859733254144

"What Is Quantum Sensing?" Bae Systems. [WEB POST]. https://www.baesystems.com/en-us/definition/what-is-quantum-sensing

"What Is Quantum-safe Cryptography?" IBM. [WEB POST]. https://www.ibm.com/topics/quantum-safe-cryptography

"WSJ: Accenture, 1QBit and Biogen POC Shows Quantum Computing May Speed Drug Discovery" 1Qbit. 2017 [PRESS RELEASE]. https://1qbit.com/news/wsj-accenture-1qbit-biogen-poc-indicates-quantum-computing-may-speed-drug-discovery/

"Xanadu and Rolls-Royce to Build Quantum Computing Tools with PennyLane" Yahoo! Finance. 2023 [NEWS POST]. https://finance.yahoo.com/news/xanadu-rolls-royce-build-quantum-130000758.html

Index

Note: **Bold** page numbers refer to tables and *italic* page numbers refer to figures.

Printed in the United States
by Baker & Taylor Publisher Services